International Association of Geodesy Symposia

Jeffrey T. Freymueller, Series Editor
Laura Sánchez, Assistant Editor

Series Editor
Jeffrey T. Freymueller
Endowed Chair for Geology of the Solid Earth
Department of Earth and Environmental Sciences
Michigan State University
East Lansing, MI, USA

Assistant Editor
Laura Sánchez
Technische Universität München
Deutsches Geodäisches Forschungsinstitut
Munich, Germany

International Association of Geodesy Symposia

Jeffrey T. Freymueller, Series Editor
Laura Sánchez, Assistant Editor

Symposium 109: Permanent Satellite Tracking Networks for Geodesy and Geodynamics
Symposium 110: From Mars to Greenland: Charting Gravity with Space and Airborne Instruments
Symposium 111: Recent Geodetic and Gravimetric Research in Latin America
Symposium 112: Geodesy and Physics of the Earth: Geodetic Contributions to Geodynamics
Symposium 113: Gravity and Geoid
Symposium 114: Geodetic Theory Today
Symposium 115: GPS Trends in Precise Terrestrial, Airborne, and Spaceborne Applications
Symposium 116: Global Gravity Field and Its Temporal Variations
Symposium 117: Gravity, Geoid and Marine Geodesy
Symposium 118: Advances in Positioning and Reference Frames
Symposium 119: Geodesy on the Move
Symposium 120: Towards an Integrated Global Geodetic Observation System (IGGOS)
Symposium 121: Geodesy Beyond 2000: The Challenges of the First Decade
Symposium 122: IV Hotine-Marussi Symposium on Mathematical Geodesy
Symposium 123: Gravity, Geoid and Geodynamics 2000
Symposium 124: Vertical Reference Systems
Symposium 125: Vistas for Geodesy in the New Millennium
Symposium 126: Satellite Altimetry for Geodesy, Geophysics and Oceanography
Symposium 127: V Hotine-Marussi Symposium on Mathematical Geodesy
Symposium 128: A Window on the Future of Geodesy
Symposium 129: Gravity, Geoid and Space Missions
Symposium 130: Dynamic Planet - Monitoring and Understanding...
Symposium 131: Geodetic Deformation Monitoring: From Geophysical to Engineering Roles
Symposium 132: VI Hotine-Marussi Symposium on Theoretical and Computational Geodesy
Symposium 133: Observing our Changing Earth
Symposium 134: Geodetic Reference Frames
Symposium 135: Gravity, Geoid and Earth Observation
Symposium 136: Geodesy for Planet Earth
Symposium 137: VII Hotine-Marussi Symposium on Mathematical Geodesy
Symposium 138: Reference Frames for Applications in Geosciences
Symposium 139: Earth on the Edge: Science for a sustainable Planet
Symposium 140: The 1st International Workshop on the Quality of Geodetic
Observation and Monitoring Systems (GuQOMS'11)
Symposium 141: Gravity, Geoid and Height systems (GGHS2012)
Symposium 142: VIII Hotine-Marussi Symposium on Mathematical Geodesy
Symposium 143: Scientific Assembly of the International Association of Geodesy, 150 Years
Symposium 144: 3rd International Gravity Field Service (IGFS)
Symposium 145: International Symposium on Geodesy for Earthquake and Natural Hazards (GENAH)
Symposium 146: Reference Frames for Applications in Geosciences (REFAG2014)
Symposium 147: Earth and Environmental Sciences for Future Generations
Symposium 148: Gravity, Geoid and Height Systems 2016 (GGHS2016)
Symposium 149: Advancing Geodesy in a Changing World
Symposium 150: Fiducial Reference Measurements for Altimetry

More information about this series at http://www.springer.com/series/1345

Fiducial Reference Measurements for Altimetry

Proceedings of the International Review Workshop on Satellite Altimetry Cal/Val Activities and Applications

Edited by

Stelios P. Mertikas, Roland Pail

Springer

Volume Editors
Stelios P. Mertikas
Laboratory of Geodesy and Geomatics Engineering
School of Mineral Resources Engineering
Technical University of Crete
Chania, Greece

Roland Pail
Institute of Astronomical and Physical Geodesy
Technische Universität München
Munich, Germany

Series Editor
Jeffrey T. Freymueller
Endowed Chair for Geology of the Solid Earth
Department of Earth and Environmental Sciences
Michigan State University
East Lansing, MI, USA

Assistant Editor
Laura Sánchez
Technische Universität München
Deutsches Geodätisches Forschungsinstitut
Munich, Germany

ISSN 0939-9585 ISSN 2197-9359 (electronic)
International Association of Geodesy Symposia
ISBN 978-3-030-39437-0 ISBN 978-3-030-39438-7 (eBook)
https://doi.org/10.1007/978-3-030-39438-7

This Springer imprint is published by the registered company Springer Nature Switzerland AG
The registered company address is: Gewerbestrasse 11, 6330 Cham, Switzerland

Preface

These proceedings include written versions of selected papers presented at the *International Review Workshop on Satellite Altimetry Calibration/Validation (Cal/Val) Activities and Applications*, held in Chania, Crete, Greece, 23–26 April 2018, organized by the Technical University of Crete, Greece. It was arranged in the context of the European Space Agency Project of "Fiducial Reference Measurements for Altimetry", and it was cosponsored by the International Association of Geodesy (in particular by the IAG Commission 2, Gravity Field), the European Space Agency, the European Union (The Copernicus Programme), the European Organization for the Exploitation of Meteorological Satellites (EUMETSAT), Space Geomatica P.C., and the Municipality of Chania. It took place at the beautiful premises of the Great Arsenali, Centre of Mediterranean Architecture, right in the middle of the Venetian old harbour of the Chania city.

The review conveners and the session chairs decided on the acceptance of the submitted abstracts. The session chairs took an active role in the selection of papers for oral and poster presentations. In addition, they organized along with IAG the review process for the papers presented in this Volume of Proceedings. The submitted papers were thoroughly peer-reviewed by a panel of international experts in the field.

The aim of the workshop was to present the latest research results in the field of satellite altimetry calibration and altimetry applications for monitoring ocean changes and for improving Earth observation in an objective, continuous, homogeneous, and reliable manner. Particular emphasis was given to understanding how to minimize uncertainties (both random and systematic) and to link complementary altimetry missions to each other. The intention has been to support the long-term monitoring of climate change by understanding better environmental changes not only on world's oceans and terrestrial surface waters, but also on Arctic and Antarctic Polar Regions. The main outcome is the establishment and promotion of a scientific roadmap with procedures, protocols, guidelines, and best practices in an open and transparent way to be followed by any international group working on satellite altimetry to attain SI (Système International d'Unités) traceability of their measurements, results, and data products.

The meeting attracted more than 70 scientists from 17 countries (Australia, Canada, China, Denmark, Estonia, France, Germany, Greece, Hungary, India, Italy, the Netherlands, Portugal, Spain, Taiwan, United Kingdom, and the USA). Major space agency and international organization representatives from the European Space Agency, EUMETSAT, NASA/Jet Propulsion Lab, Indian Space Research Organization, Centre National D'Etudes Spatiales (France), Danish Space Center, International Association of Geodesy, metrology institutes, and the International Union of Geodesy and Geophysics worked together to create a roadmap for the objective and long-term calibration of present and future satellite altimeter missions, including Sentinel-3, CryoSat-2, Sentinel-6/Jason-CS, Jason, SWOT, HY-2, and others.

The scientific program was organized by the review workshop conveners and the chairpersons of each session. Forty-six oral presentations, twenty poster presentations, and a round-table discussion on the conclusions of this International Cal/Val Review took place. Sessions were organized as follows:

- *Session 1: Fiducial Reference Measurements for Altimetry and Metrology.*

Chairs: Craig Donlon (ESA, Netherlands), Stelios Mertikas (Technical University of Crete, Greece)

- *Session 2 A: Past and Current International Cal/Val Activities.*

Chairs: Bruce Haines (JPL, NASA, USA), Pascal Bonnefond (Observatoire de Paris, France)

- *Session 2 B: Past and Current International Cal/Val Activities.*

Chairs: Mingsen Lin (NSOAS, China), Christopher Watson (University of Tasmania, Australia)

- *Session 2 C: Past and Current International Cal/Val Activities.*

Chairs: Christopher Watson (University of Tasmania, Australia), Denise Dettmering (Technical University of Munich, Germany)

- *Session 3: Evaluating Uncertainties with Metrology Standards.*

Chairs: Aloke K. Mathur (Indian Space Research Organization, India), Ambrus Kenyeres (Satellite Geodetic Observatory, Hungary)

- *Session 4: Calibration of Future Satellite Altimetry.*

Chairs: Lee Lueng Fu (JPL, NASA, United States of America), Pierre Femenias (ESA, Italy)

- *Session 5: Maintaining the Earth Observation Climate Record from Altimetry and ESA Climate Change Initiative.*

Chairs: Robert Cullen (ESA, Netherlands), Lee Lueng Fu (JPL, NASA, USA)

- *Session 6: The Changing Environment.*

Chairs: Roland Pail (Technical University of Munich, Germany), Xiaoli Deng (University of Newcastle, Australia), Michael Sideris (IUGG, Canada)

- *Session 7: Polar Regions Applications.*

Chairs: Lars Stenseng, Petr Knudsen, Ole Andersen (Danish Space Center, Denmark)

- *Sessions 8 and 9: Modelling with Altimetry: Bathymetry, Geoid, Sea Level, Gravity, Height, and Global Geodetic Observing System.*

Chairs: Ole Andersen (Danish Space Center, Denmark), Denise Dettmering (Technical University of Munich, Germany)

- *Closing Session: International Cal/Val Review Summary and Conclusions.*

Chairs: Rob Cullen (ESA, Netherlands), Stelios Mertikas (Technical University of Crete, Greece)

On the last day of the meeting, the following recommendations and actions were agreed in plenary:

- To define requirements, establish standards, and provide recommendations and best practices for altimetry calibration such that all measurements and results made for monitoring the Earth, the environment, and sea and water levels are well characterized and linked to the *International System of Units* (SI) and *Metrology Standards*;
- To document procedures for *fiducial reference measurements* in satellite altimetry calibration so that results are traceable to SI units, reliable in the long-term, and comparable worldwide, to support an objective and unquestionable monitoring of sea level and climate change;
- To establish procedures and protocols for characterizing *uncertainty budgets* of all fiducial reference measurement instruments and derived results over the entire end-to-end duration of a satellite mission in order to support a rigorous treatment and a trustworthy assessment for uncertainties (error constituents, properties, relationships between them, and so on) in altimetry calibration;
- To define and document procedures, protocols, and best practices to evaluate differences in instrument performances under a range of conditions (frequency of passes per altimeter at Ku and Ka-bands, sensor replacement, geographical distribution, latency, etc.) to support Earth observation by altimetry for the future;

- To adopt a *standard time* and *spatial reference frame* for all in-situ measurements, used in the calibration and validation of space-borne altimeter systems;
- To establish a harmonized approach to in-situ performance, functionality, and availability;
- To have an *operational* and *continuous* capability to support a minimum set of required observations for altimetry Cal/Val;
- To establish a consolidated approach to data formatting, archiving, and distribution;
- To provide proper and good quality documentation and practical guidelines not only for those unfamiliar with the altimetry data, but also for the data providers to better manage data production, storage, updating, and reuse.

If all the above are implemented properly, the strategy of fiducial reference measurements for altimetry will provide the maximum return on investment for a satellite mission. It will also furnish to users the required confidence in data products, in the form of independent validation results and satellite measurement uncertainty estimation, over the entire end-to-end duration of a satellite mission.

The scientific committee of this International Cal/Val Review consisted of Stelios Mertikas, (Technical University of Crete, Greece), Craig Donlon (European Space Agency, The Netherlands), Erik de Witte (European Space Agency, The Netherlands), Robert Cullen, (European Space Agency, The Netherlands), Pierre Féménias (European Space Agency, Italy), Constantin Mavrocordatos (European Space Agency, The Netherlands), Jérôme Benveniste (European Space Agency, Italy), Remko Scharroo (EUMETSAT, Germany), Ole B. Andersen, (Danish Technical University, Denmark), Pascal Bonnefond (L'Observatoire de Paris, SYRTE, France), Francois Boy (Centre National d'Etudes Spatiales, France), Jean-Francois Crétaux, (LEGOS, France), Xiaoli Deng (Newcastle University, Australia), Nigel Fox (National Physical Laboratory, UK), Bruce Haines (Jet Propulsion Lab/NASA, USA), Osamu Isoguchi (Remote Sensing Technology Center of Japan, Japan), Ambrus Kenyeres (Institute of Geodesy Cartography and Remote Sensing, FOMI, Hungary), Ronald Kwok (Jet Propulsion Lab/NASA, USA), Lee-Lueng Fu (Jet Propulsion Lab/NASA, USA), Mingsen Lin (National Ocean Satellite Application Center, China), Aloke K Mathur (Indian Space Research Organization, India), Daniel Medeiros Moreira (Geological Survey of Brazil CPRM, Brazil), Roland Pail (Technical University of Munich, Germany), Ilias N. Tziavos (Aristotle University of Thessaloniki, Greece), and Christopher Watson (University of Tasmania, Australia).

The local organizing committee consisted of Stelios P. Mertikas (Technical University of Crete), Achilles Tripolitsiotis (Space Geomatica), Stella Galani (Space Geomatica), Dionissis Efstathiou (Technical University of Crete), Constantine Kokolakis (Technical University of Crete), Demitris Galanakis (Space Geomatica), and Xenofon Frantzis (Technical University of Crete).

The IAG, ESA, and EUMETSAT approved several travel awards for participants. Financial support and promotional support was given by a number of agencies. Special thanks go to the International Association of Geodesy, the Technical University of Crete, the European Space Agency, and EUMETSAT.

To all individuals who have, in one way or another, been involved in the preparation of the *International Review Workshop on Satellite Altimetry Calibration/Validation (Cal/Val) Activities and Applications* and to all organizations, Session Chairs, Reviewers, and Committees that have given their support, I extend sincere thanks. It is their hard work and support that laid the foundation for the success of this event. In addition to those cited above for their assistance, I am grateful to the ESA mission scientist, Craig Donlon, for his continual support, encouragement, and inspiration. I am also thankful to Josef Aschbacher, Director of Earth Observation Programmes at the European Space Agency, Alain Ratier, Director-General at EUMETSAT, the President of the IUGG, Michael G. Sideris, the IAG Secretary General, Hermann Drewes, and the rector of the University, Evan Diamantopoulos, for their help and support.

A final word of thanks goes to Jeffrey T. Freymueller (Michigan State University, USA) and Laura Sánchez (Technical University of Munich, Germany) who coordinated, on behalf of the IAG, the review process and Roland Pail (Technical University of Munich, Germany) who acted as a volume editor and strongly support the paper review process.

Chania, Greece
14 September 2019

Stelios P. Mertikas

FRM4ALT

Contents

Fiducial Reference Measurements for Satellite Altimetry Calibration: The Constituents

Stelios P. Mertikas, Craig Donlon, Pierre Féménias, Rob Cullen, Demitris Galanakis, Xenophon Frantzis, and Achilles Tripolitsiotis

Abstract

This work defines the concept of Fiducial Reference Measurements (FRM) for altimetry calibration. It has emerged out of the requirement for reliable, consistent, standardized and undisputable Earth observation records. FRM observations are to become comparable worldwide, insensitive to instrument, satellite, setting, location and measurement conditions among others. At first, the paper singles out the various contributing error sources to the calibration error budget in altimetry. Secondly, it evaluates measurement uncertainty as it originates from each constituent in the error propagation chain. Third, it sets the foundation for connecting measurement uncertainty so that it could be traceable to international metrology standards (speed of light, atomic time, for example). Finally, this paper presents procedures, protocols and best practices for arriving at FRM standards in sea-surface but also in transponder calibration. This analysis is based on experience gained over the last 15 years of operation of the permanent facility for altimetry calibration in west Crete, Greece.

Keywords

Altimetry · Calibration · Error budget · Fiducial reference measurement

1 Introduction

Even small amounts in sea-level rise, measured at +3.2 mm/year today with altimetry, can cause devastating effects. Sea level rise causes destructive erosion on coasts, contaminates faster aquifers and water resources with sea water (e.g., Messara valley in Crete, Greece, Paleologos and Mertikas 2013), but also harms agriculture and productive soils. For every 30 cm of sea level rise, coastlines move inland by 30–100 m on average (Meltzner et al. 2017). Coastal flooding destroys wildlife (i.e., birds, fish, animals, vegetation), but also sea level rise causes hurricane surges to become powerful, higher, with frequent flooding on vast coastal areas. In the near future, islands may be lost and people living on low-lying lands may abandon their homes and relocate.

Sea level monitoring requires longstanding observations of several decades; a duration going beyond the typical observation length and lifetime for a satellite altimeter of about 5–7 years. In addition, the sound identification of sea-level signals connected to long-term climate changes still remains a challenge, although contemporary altimeters provide remarkable accuracies (Fu and Haines 2013). Consequently, to support a seamless, reliable and objective monitoring for sea level and inland waters, tied to an inertial reference system (Müller 2014), diverse satellite altimeters

S. P. Mertikas (✉) · X. Frantzis
Geodesy and Geomatics Engineering Lab, Technical University of Crete, Chania, Crete, Greece
e-mail: mertikas@mred.tuc.gr

C. Donlon · R. Cullen
ESA/ESTEC, Noordwijk, ZH, The Netherlands

P. Féménias
ESA/ESRIN, Frascati, Italy

D. Galanakis · A. Tripolitsiotis
Space Geomatica P.C., Chania, Crete, Greece

S. P. Mertikas, R. Pail (eds.), *Fiducial Reference Measurements for Altimetry*, International Association of Geodesy Symposia 150, https://doi.org/10.1007/1345_2019_56

have to be calibrated but also cross-calibrated against each other continuously and in an absolute sense by ground-truth research infrastructures (Parkinson 2017).

Absolute calibration of satellite altimeters by external and independent to satellite observations is a prerequisite for a continuous, homogenous and reliable monitoring of the Earth, its oceans and climate change. These calibration/validation (Cal/Val) facilities on the Earth's surface ensure that altimetry observations are free of errors and biases, uninterrupted, but also tied from one mission to the next in an objective and absolute sense. Altimeter system's responses have to be, thus, continuously monitored and controlled for their quality, biases, errors, drifts, and the rest. Relations also among different missions have to be established on a common and reliable Earth-center reference system, maintained over a long period of time.

Ground calibration and validation facilities are located either offshore or on land (e.g., microwave transponders) exactly under (absolute direct Cal/Val) or adjacent to satellite's ground track on nearby coasts, to ensure monitoring of uncontaminated satellite observations (absolute indirect Cal/Val). Relative calibration of satellite measurements is also performed using either multi-mission crossover analysis between reference altimeters and other missions (relative direct Cal/Val) or distributed tide-gauge networks (relative indirect Cal/Val).

The international altimetry community expects continuity and upscaling of Cal/Val services to maintain measurement conformity and error reporting, but also to support the right decisions concerning Earth observation. With the recent advent of diverse satellite altimeters along with advanced measuring techniques (Nadir, Delay-Doppler, interferometry altimeters, wide swath, Ku-band, Ka-band frequencies), it has become mature and is high time to maintain absolute reference Cal/Val sites to regularly monitor any altimeter. However, monitoring should be based upon fundamental and undisputable reference and metrology standards, i.e., speed of light, absolute time. This concept of "Fiducial Reference Measurement (FRM) for altimetry" (Donlon 2018; Loew et al. 2017) has been introduced by the European Space Agency (ESA) and for the future, ESA plans to calibrate all its present and forthcoming altimeters in that manner.

These FRM Cal/Val sites will constitute the fundamental mainstay for building up capacity for monitoring climate change in an objective and unequivocal manner with altimetry. They will be capable of assessing any altimeter measurements to absolute reference signals traceable to SI-standards (Système International d'Unités, such as speed of light, absolute time reference, BIPM (2019)) with different techniques, various processes and diverse instrumentation and settings.

This paper delineates major error contributors in sea-surface but also in transponder calibration. Error constituents are identified and a few recommendations for expressing uncertainties as well as ways for arriving at FRM standards in altimetry calibration are given. This analysis has been based on experience gained over the last 15 years of operation of the permanent facility for altimetry calibration in Gavdos and west Crete, Greece (Mertikas et al. 2018a, b).

2 Constituents in Sea-Surface Calibration

Five main groups of constituents are responsible for the final results in sea-surface calibration. These are: (1) Absolute coordinate determination, (2) Sea-surface height, (3) Control ties and settings, (4) Height transfer from coastal sea surface to open sea, and (5) Processing errors. Each constituent carries its own innate uncertainty which is finally transferred and propagated to the ultimate error estimation in the results of altimetry calibration. Procedures for sea-surface calibration will not be presented here as these have already been reported elsewhere (Mertikas et al. 2010).

2.1 Absolute Coordinates for the Cal/Val Site

The contributing sources of uncertainty to the final coordinate determination for the Cal/Val site originate from diverse components and influences, indicatively such as:

1. *Site Location*. Influencing factors include type of harbor and its stability, ground deformation, seismic activity, surrounding seas and conditions (i.e., river runoffs, ocean currents, local dynamics and effects) geoid uncertainty, measuring conditions, satellite visibility, proximity to altimeter ground tracks, tide gauges, and so forth. An extensive description of the Gavdos/Crete Cal/Val facilities is given in Mertikas et al. (2018a, c).
2. *GNSS Instrumentation*. Uncertainties may arise from antenna and receiver type, redundant receivers and positioning systems (diverse systems, multi-frequency, multi-constellation, sampling, and so forth), processing strategies, reference points and geodetic ties, sampling rates, measurement type, and so forth. For example, each GNSS antenna at the Cal/Val site shall be characterized in specialized labs and uncertainties should not entirely depend upon those provided by either the International GNSS Service or the manufacturer. Differences of several mm have been discovered in such cases (Mertikas et al. 2018b). At least 2–3 years of continuous GNSS operations are needed to claim uncertainties of $<\pm 1$ mm in absolute coordinate determination.
3. *Reference Systems*. Coordinates shall be reported to an international reference system (i.e., GRS80 ellipsoid,

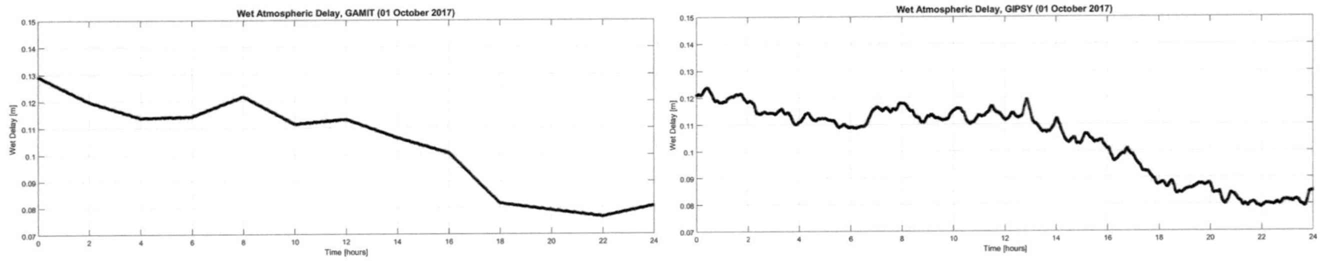

Fig. 1 Zenith Wet Delay as estimated by GNSS processing using differential (left diagram) and precise point positioning (right diagram) for 1-Oct-2017

ITRF2014, WGS-84, PZ-90, Galileo TRF, BeiDou reference system) with known transformations to altimetry products. Uncertainties may emerge for example from applied transformations, control points, relations to reference surfaces, and the like.

4. Earth Tides. Earth tides (solid, loading, pole) influence site coordinates at the time of satellite pass. These are essential constituents and should be included when trying to achieve FRM status for altimetry calibration. Tide prediction models and associated uncertainties depend upon location, tide magnitude, bathymetry and other loading factors.
5. *Atmospheric Delays.* Troposphere and ionosphere delays impact final positioning of the Cal/Val site. The wet troposphere delays present spatial and fast temporal variations so their uncertainty is ambivalent and hard to estimate and incorporated into the GNSS processing. Different scientific GNSS processing employ diverse estimations for the zenith wet troposphere delays (Fig. 1).
6. Time Reference for GNSS Observations. The diversity on reference time systems applied by different GNSS systems (i.e., GPS: GPS time (USA), DORIS: TAI (Doris), GLONASS: UTC (Soviet Union), Galileo: Galileo System Time (European), BeiDou: BeiDou Time (China)) and its uncertainty has to be accessed first and fed into the error budget of Cal/Val results.

The target for absolute altimeter calibration is to achieve positioning accuracies for the Cal/Val site of the order of $<\pm 1$ mm and with respect to the center of mass of the Earth.

2.2 Water Level Determination at the Cal/Val Site

Several sources of uncertainty are identified when the water level is determined at a Cal/Val site whereas satellite calibration is carried out in open sea. A few of those are singled out and outlined below:

1. *Site Location.* The site has to be as close to the satellite ground track as possible, impervious to any human activities, accessible, protected against waves, local sea level effects, but at the same time, be at a location to sense and feel the open sea conditions and dynamics. Uncertainties related to site effects (local conditions and harbor dynamics, bathymetry and atmospheric loading along with others) have to be evaluated as they contribute to errors for the water level determination.
2. *Conditions and Settings.* Environmental conditions influence water level determination. Thermal expansion and wind loading on supporting structures of tide gauges, high frequency and inverse barometer variations but also differences in sea level between stilling well, harbor and outside of it, along with others, are responsible for extra uncertainties in water level determination.
3. *Water Level Measuring Sensors.* Different types of instrumentation (i.e., acoustic, radar, pressure, floating) are available to determine water level at the time of satellite pass. Measuring and data logging strategies, zero-reference measuring points, sampling rates, estimation procedures, along with others, are associated with an uncertainty for water level determination. All these contributions have to be continuously monitored for offsets, drifts, and other effects. Master tide gauges could set the standards for uncertainty evaluations and alterations from it for all other tide gauges on site. Regular (i.e., semi-annual) validation experiments with tide pole readings are carried out for that purpose (Fig. 2).

A final value for water level could be estimated as a conglomerate mean (e.g., averaging, filtering, interpolation) of all water level sensors along with their measurement uncertainty before allowing these values to take part in Cal/Val processing.

The aim of absolute altimeter calibration is to achieve water level uncertainties of less than 0.5 cm and with respect to the center of mass of the Earth.

2.3 Control Ties and Settings

The objective for this activity is to set the foundation for monitoring any changes of the ground supporting the Cal/Val

Fig. 2 Field calibration of tide gauges in Gavdos Cal/Val, Greece. A digital camera placed in front of a tide pole to capture actual sea surface level can be used to calibrate the operating tide gauges

site, to make provisions for securing and recovering the location in case of damage but also to connect the GNSS and water level measurements. A number (at least five) of control ties (i.e., benchmarks, geodetic control points) shall be established around the site (<100 m) to secure fundamental reference points used for calibration. The height difference between these benchmarks, the GNSS antenna reference point and the tide gauges zero-measuring point shall be determined with uncertainties of ± 1 mm but also monitored at least every 6 months. Differences between a tide gauge and a GNSS reference point, as well as uncertainties arising from spirit levelling, geoid heights as well as from models for Mean Sea Surface and Mean Dynamic Topography provide additional components for error estimation in water level determination at the Cal/Val site.

2.4 Height Transfer from Cal/Val Site to Open Sea

The sea-surface height as established at the Cal/Val site has to be transferred to the open sea where calibration takes place with uncontaminated satellite measurements. This height transfer is dependent upon local and regional reference surfaces and models (i.e., mean sea surface, mean dynamic topography and geoid among other things). These models and surfaces have to be verified by external means and field campaigns (e.g., GNSS boat campaigns, gliders). Principally, reference calibration models hinge on the Cal/Val region characteristics and peculiarities and carry their uncertainty for height transfer.

2.5 Processing for Sea-Surface Calibration

Any procedure for satellite altimetry calibration (e.g., satellite orbit, altimetry, tide gauge, GNSS, reference model among other contributions) brings along its associated processing uncertainty. The algorithms for example to interpolate, extrapolate, filter, average, and so forth, the raw measurements and results come with errors which have to be taken into account in the overall uncertainty estimation for processing.

3 Constituents in Transponder Calibration

The uncertainty associated with the altimeter calibration using a transponder is broken down to the following contributing factors:

3.1 Cal/Val Site Coordinate Determination

This element is similar to the one previously described for sea-surface calibration. Its only difference lies on the estimation of wet troposphere delays at the transponder Cal/Val site, as satellite altimeter radiometers are primarily inoperative over land. Thus, other techniques for estimating wet troposphere delays at the Cal/Val site of the transponder have to be employed (radiometers, GNSS-derived delays, numerical weather models, radiosondes). These carry their own uncertainty, and in addition to other uncertainties previously outlined in Sect. 2.1.

3.2 Time Reference

Time reference for transponder calibration involves important issues. These can be broken down to the following constituents: (1) When distances are measured from the satellite to the transponder, a question comes up as to what "distance" is defined and by which time standard (satellite time, terrestrial time, dynamic time, along with others), but (2) what should be the correct term for the parameter "distance" to calibrate with the transponder if we had available an absolute time standard operating at the Cal/Val site. This is also true for defining the internal delay of the transponder itself. All of the above are associated with uncertainty budget to be taken into consideration for transponder calibration.

3.3 Control Ties and Settings

This constituent is identical to the one presented in previous Sect. 2 in Sea-Surface Calibration.

3.4 Transponder Data Processing

Several uncertainty parameters are involved in transponder calibration. These are related to: (a) satellite's position and velocity over the transponder, (b) the phase center of the transmitting satellite antenna and its variations as a function of pitch, yaw and roll angles of the satellite's orientation; (c) the reference bin number of the returned pulse, (d) the implemented Digital Elevation Models (Doppler effects), (e) parameters influencing the signal tracking and range measurement by the satellite, and so on. Each of these constituents brings along its own uncertainty which in turn is propagated into the final error budget for transponder calibration.

Both activities for sea-surface and transponder uncertainty analysis involve a category of unaccounted effects. This group of uncertainty constituent is to include any additional effects (altimeter instruments, orbit, for example.) that have not taken care of in the previous descriptions.

4 Conclusions

The sea-surface and transponder altimetry calibrations have been broken down to the factors which influence their overall uncertainty. This is the first step towards the realization of fiducial reference measurements for satellite altimetry calibration. The following general guidelines are recommended when assessing measurements uncertainties in FRM altimetry calibration: (1) *Measurement conformity*. All measurements and data for each contributing component in altimetry calibration have to be thoroughly inspected for gross errors (Barnett and Lewis 1994), stochastic and deterministic structures, measurement irregularities, invalidity of models, symmetry and serial correlation among other things, before they are fed into altimetry calibration chain. (2) *Decision making standards*. Choose the measurand to represent "uncertainty" as the relevant carrier of true information for expressing accuracy in Cal/Val results. The selected mathematical and quality tool to express uncertainty have to be complete (represent all diversities and nuances in Cal/Val), efficient (small variance), reliable (no matter what the statistical distribution) and stable (same results on the same Cal/Val site). (3) *Weigh each contributing source*. Assess the conventional but also the "robust" uncertainty of each contributing source. Individual weights and sensitivity factors for each component will come out of this evaluation. And (4) *Uncertainty reporting*. Integrate each weighted individual contribution along with its sensitivity factor to produce the final report for the uncertainty in altimetry calibration results.

Examples of the FRM uncertainty budget analysis for sea-surface calibration has been presented in Mertikas et al. (2018a) and for the transponder calibration in Mertikas et al. (2018c).

References

Barnett V, Lewis T (1994) Outliers in statistical data, 3rd edn. Wiley, New York

BIPM (2019) The International system of units (SI), 9th edn. Bureau International des Poids et Measures, Paris. https://www.bipm.org/cc/CCU/Allowed/23/Draft-SI-Brochure-2018.pdf. Accessed 10 Jan 2019

Donlon C (2018) Fiducial reference measurements for altimetry. FRM4ALT Project Webportal. https://goo.gl/Yn23pQ. Accessed 26 June 2018

Fu LL, Haines BJ (2013) The challenges in long-term altimetry calibration for addressing the problem of global sea level change. J Adv Space Res 51:1284–1300. https://doi.org/10.1016/j.asr.2012.06.005

Loew A, Bell W, Brocca L, Bulgin CE, Burdanowitz J, Calbet X, Donner RV, Ghent D, Gruber A, Kaminski T, Kinzel J, Klepp C, Lambert J-C, Schaepman-Strub G, Schröder M, Verhoelst T (2017) Validation practices for satellite-based Earth observation data across communities. Rev Geophys 55:779–817. https://doi.org/10.1002/2017RG000562

Meltzner AJ, Switzer AD, Horton BP, Ashe E, Qiu Q, Hill DF, Bradley SL, Kopp RE, Hill EM, Majewski JM, Natawidjaja DH, Suwargadi BW (2017) Half-metre sea-level fluctuations on centennial timescales from mid-Holocene corals of Southeast Asia. Nat Commun 8:14387. https://doi.org/10.1038/ncomms14387

Mertikas SP, Ioannides RT, Tziavos IN, Vergos GS, Hausleitner W, Frantzis X, Tripolitsiotis A, Partsinevelos P, Andrikopoulos D (2010) Statistical models and latest results in the determination of the absolute bias for the Radar Altimeter of Jason satellites using the Gavdos facility. Marine Geodesy 33(1):114–149. https://doi.org/10.1080/01490419.2010.488973

Mertikas S, Donlon C, Féménias P, Mavrocordatos C, Galanakis D, Tripolitsiotis A, Frantzis X, Tziavos IN, Vergos G, Guinle T (2018a) Fifteen years of Cal/Val service to reference altimetry missions: calibration of satellite altimetry at the permanent facilities in Gavdos and Crete, Greece. Remote Sens 10:1557. https://doi.org/10.3390/rs10101557

Mertikas S, Donlon C, Féménias P, Galanakis D, Tripolitsiotis A, Frantzis X (2018b) International standardization for satellite altimetry calibration: lessons for the past & roadmap to the future. In: Ocean Surface Topography Science Team Meeting 2018, 24-29 Sept. 2018, 24-29 Sept. 2018, Ponta Delgada

Mertikas S, Donlon C, Féménias P, Mavrocordatos C, Galanakis D, Tripolitsiotis A, Frantzis X, Kokolakis C, Tziavos IN, Vergos G, Guinle T (2018c) Absolute calibration of the European Sentinel-3A surface topography mission over the permanent facility for Altimetry Calibration in west Crete, Greece. Remote Sens 10(11):1808. https://doi.org/10.3390/rs10111808

Müller R (2014) Calibration and verification of remote sensing instruments and observations. Remote Sens 6:5692–5695. https://doi.org/10.3390/rs6065692

Paleologos EK, Mertikas SP (2013) Evidence and implications of extensive groundwater overdraft-induced land subsidence in Greece. Eur Water 43:3–11

Parkinson C (2017) Satellite contributions to climate change studies. Proc Am Philos Soc 161(3):208–225

Practical Options for Time Tagging

Demetrios Matsakis

Abstract

The international timekeeping community prepares several official timescales, and there exist a plethora of alternatives used by others, often with no official recognition. The advantages and disadvantage of several official timescales are presented; however, they are all interconvertible using information readily available on the internet. A discussion of traceability and the statistical errors is included; all of these scales are ultimately traceable to UTC. Many of the considerations are not relevant if the error tolerance is above a few tens of nanoseconds.

Keywords

GNSS · GPS · TAI · Time-tagging · Traceability · UTC

1 TAI and UTC

The International Bureau of Weights and Measures (BIPM), under the auspices of the Treaty of the Meter, produces Coordinated Universal Time (UTC) and International Atomic Time (TAI) on a monthly basis, with a 5-day granularity and a time lag that is typically around 10 days. The difference between them is that UTC has pre-announced one-second jumps (leap seconds) that are currently inserted roughly every 2 years. In most technologically developed nations UTC is the basis of legal time standard, and their national laboratories provide a real-time realization. For example, the U.S. Naval Observatory's (USNO) Master Clock's time is termed UTC (USNO), and this is the timing reference for GPS.

The motivation for leap seconds in UTC is to allow for the variable and unpredictable variations in the rotation of the Earth (Fig. 1), which in the long term slows down due to tidal friction. Over decadal scales the Earth can speed up, and over annual and sub-annual scales it is strongly anticorrelated with the atmospheric angular momentum (A discussion of variations in the Earth's rotation can be found in the "Science Background", n.d.).

In Fig. 1 (right), the discontinuities are due to leap seconds that were inserted to ensure that the difference between UTC and UT1 (universal time defined by the Earth's rotational angle) is less than 0.9 s in absolute value. UT1-UTC and a list of leap seconds are published by the International Earth Rotation and Reference Service (IERS) in www.iers.org. The value of UT1-UTC will also be broadcast by planned GNSS upgrades. Table 1 summarizes how time-tagging with UTC would be affected by leap seconds. Leap seconds have historically been inserted at the end of the months of June and December, however the formal specifications allow for their insertion at the end of any month. It is a fact that every recent leap second insertion has led to failures with some hardware and software systems; this could affect the time-tagging equipment as well. In the event the Earth speeds up enough, it is possible that a second could be skipped. Although the informed user, software writer, or manufacturer can in theory design systems to be immune to the possibly disastrous effects of such UTC "discontinuities", a somewhat controversial effort is underway to eliminate leap seconds. A more thorough discussion can be found in http://tycho.usno.navy.mil/papers/ts-2014/Matsakis-LeapSecondComments.

D. Matsakis (✉)
Silver Spring, MD, USA
e-mail: dnmyiasou@yahoo.com

S. P. Mertikas, R. Pail (eds.), *Fiducial Reference Measurements for Altimetry*,
International Association of Geodesy Symposia 150, https://doi.org/10.1007/1345_2018_53

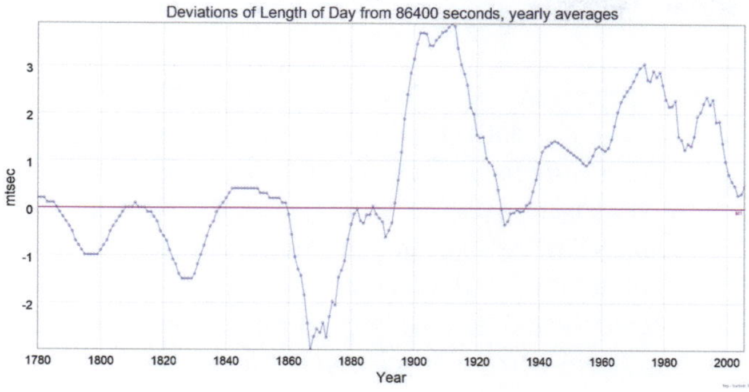

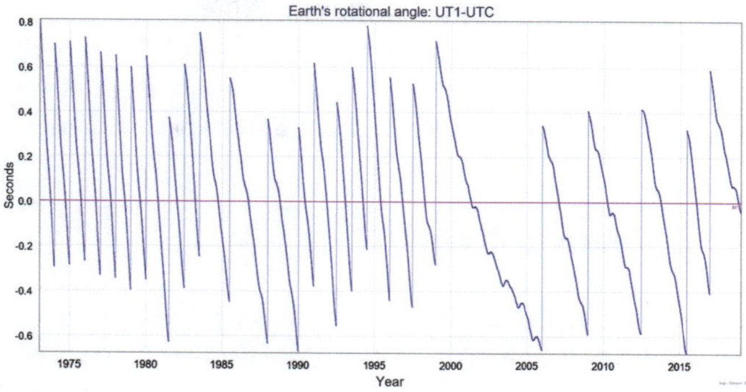

Fig. 1 The observed Length of Day (LOD) since 1760 (left), and UT1-UTC since 1973 (right). The pre-atomic clock data are based on the difference between the rotation of the Earth and the far more stable orbital motions of the Moon and planets. UT1-UTC is the integral of the (inverse) LOD after incorporating leap seconds

Table 1 Sequential order of seconds preceding, including, and following a leap second introduced at the end of a month

Calendar day in UTC	UTC HH:MM:SS	Type of second
December 31	23:59:58	Normal
December 31	23:59:59	Normal – skipped if a "negative leap second" is needed to compensate for Earth speedup
December 31	23:59:60	Leap Second – inserted when needed to compensate for Earth slowdown
January 1	00:00:00	Normal

URSI-2014.pdf and https://www.gps.gov/cgsic/meetings/2014/.

2 Relativistic Timescales

Terrestrial Time (TT) was initially defined as a coordinate time with a 32.184 s offset from TAI at its origin and with the same rate as an observer stationary on the geoid. At the 2000 General Assembly of the International Astronomical Union (IAU), the term "geoid" was replaced by a fixed frequency offset from Geocentric Coordinate Time (TCG), which is the time kept by an observer stationary with respect to the Earth center's rest frame, but far enough away that the Earth's gravity can be ignored. The BIPM produces a yearly realization of Terrestrial Time (TT) with 10-day granularity, termed TT(BIPMxx), where xx are the last two digits of the year, as an offset to TAI and UTC. This is a post-processed timescale, designed for long-term stability and to be suitable for pulsar data analysis. Because the annual re-computations may introduce new corrections, or alter old ones, time-tagging via TT is not a good idea. For analysis, data time-tagged in TAI or UTC can always be reduced to TT(BIPMxx), for the most recent year available. Using these TT values, it is then possible to relate the data to TCG or Barycentric Coordinate Time (TCB), which is referenced to the barycenter of the Solar System, using analytic formulas published in the IERS Conventions and available at maia.usno.navy.mil or at www.iers.org.

3 GPS Timescales, and Other GNSS Timescales

The GPS Master Control Station's Kalman filter does not compute GPS time directly; however, each corrected clock can be considered a representation of GPS Time. GPS time is then an implicit timescale equal to the weighted average of all corrected ground and satellite clocks in the system. GPS Time itself is intended for navigation only and therefore is not adjusted for leap seconds; its offset from UTC therefore changes with each leap second but it remains 19 s plus a few ns offset from TAI. Since each satellite's broadcast of clock and orbit corrections is based upon daily updates, their accuracies degrade by a few ns between updates. Therefore, any direct evaluation of GPS time depends on how recently the satellites that happen to be in view have been refreshed.

GPS also broadcasts, in Subframe 4, Page 18 of the navigation message, the leap second correction as well as other parameters needed to infer (predict) UTC(USNO) from GPS Time. The value of UTC(USNO)-UTC(USNO via GPS), as measured with the broadcast parameters of satellites observable at the USNO, and downloadable from ftp://tycho.usno.navy.mil/pub/gps, is less than 1 ns RMS when averaged over a day.

Several GNSS systems are rapidly attaining operational status, which will benefit all users. However, at this time the BIPM only evaluates GPS and GLONASS in its monthly evaluation. GLONASS currently incorporates leap seconds directly into its system time, so that it has only one timescale, which is a prediction of the Russian timescale UTC(SU).

GLONASS system time has in the past suffered from a large bias, however that situation has improved over the last several years. GALILEO system time is based upon an average of the UTC realizations of five participating European laboratories.

4 Traceability to UTC

Traceability is defined in the International Vocabulary of Metrology as the property of a measurement result whereby the result can be related to a reference through a documented unbroken chain of calibrations, each contributing to the measurement uncertainty (http://www.bipm.org/en/publications/guides/vim.html). This section summarizes a paper on traceability given by myself and two employees of the National Institute of Standards and Technology (NIST) at the 2018 meeting of ION-PTTI (Matsakis et al. 2018). It was our conclusion that, since the differences between UTC and all of the timescales appearing in Table 1 are computed monthly by the BIPM in the Circular T (or, for GPS, in the associated BIPM web pages), along with their uncertainties, any of them can be used for traceability if the user has taken appropriate care to calibrate the local equipment and estimate the associated uncertainties. The uncertainties of UTC-GNSS are given as 10 ns, while those of UTC-UTC(k) are usually considerably less. Unfortunately, some of the UTC(k) are based upon uncalibrated systems, and their systematic uncertainties are listed as 20 ns – in reality those uncertainties are undefined and data involving those labs should not be used for this time-tagging.

Uncertainties can be computed using the Guide to Uncertainty in Measurements (GUM), available at https://www.bipm.org/en/publications/guides/gum.html. This reference breaks the uncertainties in two types. Type A (u_A) uncertainties are statistical in nature, they can in general be improved through averaging. Type B (u_B) uncertainties are systematic and can be reduced through calibration. The total uncertainty can be written as an agreed-upon multiple N of the root-sum-square (RSS) of the uncertainty types, i.e. N $\sqrt{{u_A}^2 + {u_B}^2}$.

In combining uncertainties to form a traceability chain from the user's system to UTC or TAI, the correlation between the links of the chains should be taken into consideration (Matsakis et al. 2006). Uncorrelated links are combined using the RSS of their uncertainties. If traceability is obtained via an intermediate system, then any uncertainties that are the property of that system should be ignored because they would be anticorrelated between the entering and departing link. For example if we obtain the time difference Equipment-UTC by first measuring Equipment-UTC(k) and then UTC-UTC(k), any error or uncertainty component that is identically the same for all measurements involving UTC(k) will drop out when the two expressions involving UTC(k) are differenced.

5 Time Transfer

Although many means of time transfer are possible (Matsakis et al. 2014), it is anticipated that the timetags will be based on GNSS measurements, and hopefully supplemented with a sanity check using a parallel technique such as Network Time Protocol (NTP). As indicated in (Matsakis et al. 2014), the precision of GNSS techniques can be 1–10 ns at a day. The 10 ns upper bound is roughly the difference between using an ionosphere correction measured via the two-frequency method, and the less accurate ionosphere correction available using the GPS broadcast parameters of the Klobuchar model.

If a geodetic receiver is available, then time transfer via Precise Point Positioning (PPP) may be preferable as it is more precise and less sensitive to diurnal temperature and humidity variations. A number of free services are available that will accept uploaded RINEX (data) files generated by the receiver and after a short delay make available solutions giving the receiver's position and also its reference time compared to an International GNSS Service (IGS) Timescale. Three such providers are Jet Propulsion Laboratories (JPL), the National Resources Canada (NRC), and the Center for Orbit Determination in Europe (CODE). Many participating UTC labs also make their RINEX files available, and by using in-house software or one of these services, one can therefore obtain IGS Timescale-UTC(k), for lab (k). By differencing these files, and using the Circular T, the user equipment can be referenced to UTC. However the user must take care to consult with the laboratory that is the source of the files, as some do not include calibration values in the header data.

If PPP is not practical, then time transfer can be made using the receiver's code data in conjunction with code data that many UTC labs make available on their web sites. It would be best to first apply orbit and atmosphere corrections computed by the International GNSS Service (IGS), which are typically a few ns. In the Common View method, the code data from the user's receiver and UTC-lab's receiver are used to measure the time difference between each satellite and each receiver over intervals of a few minutes. For those intervals, the differences with the satellite are double-differenced to obtain a time difference between the labs, and the satellite clock drops out (as it is correlated between the two sites and anti-correlated in the equations that form the link between the two sites). Residual errors due to orbit or atmosphere also drop out for nearby receivers. However, for widely separated site pairs, the signal to noise falls dramatically because the number of mutually observable satellites at any instant is less. In addition, and especially if IGS orbits and atmospheres are not used to correct broadcast data, the differential errors due to those effects increase with baseline length. Under such circumstances, it becomes more advantageous to use the All-In-View mode. In this mode, the

Table 2 Summary of time-tagging options

Timescale	Continuous?	Monotonic?	Must correct for Earth rotation?	Approximate latency	Notes
TAI	Yes	Yes	Yes	1 month	
UTC	Jumps at leap seconds	Only if all software written with care	Not if 1 s error is ok	1 month	Hardware and/or software may not handle leap seconds correctly
UTC(k)	Jumps at leap seconds	Only if all software written with care	Not if 1 s error is ok	Real-time	
TT	Yes	Yes	Yes	1 year	Published as difference with UTC and TAI
TCG	Yes	Yes	Yes	Deterministic transformation of TT	Ref to time at center of Earth
TCB	Yes	Yes	Yes	Deterministic transformation of TT	Ref to solar system barycenter
GPS Time	Yes	Yes	Yes	Real-time	Navigational timescale TAI + 19 s
UTC(USNO) from GPS	Jumps at leap seconds	Only if all software written with care	Not if 1 s error is ok	Real-time	GPS Time + broadcast corrections
GLONASS Time	Jumps at leap seconds	Only if all software written with care	Not if 1 s error is ok	Real-time	System time is tied to UTC(SU)

We note that these modes are not necessarily exclusive since one could for example store both UTC and TAI with the data

data for all satellites observable at each site at any instant are first averaged, and then the averages from the two sites are differenced. PPP can be considered a form of All-in-View that uses both code and phase data.

6 Calibration

The most difficult part of estimating the uncertainties will likely be at the user end. The calibration of equipment can be altered by temperature, humidity, vibrations, aging, power cycling, apparently spontaneous and often not-immediately-noticeable component failures, reflections, and impedance mismatches (Jiang et al. 2017). Simply inserting cables can change the calibration non-linearly. Systematic multipath can affect GNSS timing. Equipment supplied by manufacturers would best be verified by the user; the most erroneous GPS receiver calibration in our experience was off by 700 ns. Often these effects can be combined, or intermittently masked, in not-immediately-obvious and not-exactly-repeating manners. The history of science has many examples of serious misunderstandings caused by inaccurate calibration.

7 Conclusions

The importance of calibration is again emphasized, and a summary of the time-tagging options is presented in Table 2.

Disclaimer

The ideas and opinions expressed here are entirely the author's and not necessarily those of his employer, the U.S. Department of Defense, or the U.S. government.

References

A discussion of variations in the Earth's rotation can be found in the "Science Background" tab of www.iers.org

Jiang Z, Matsakis D, Zhang V (2017) Long-term instability in UTC time links. ION-PTTI

Matsakis D, Levine J, Lombardi MA (2018) The metrological and legal traceability of time signals. ION-PTTI. https://tf.nist.gov/general/pdf/2941.pdf and http://tycho.usno.navy.mil/papers/ts-2018/Traceability-of-Time-Signals.pdf

Matsakis D, Arias F, Bauch A, Davis J, Gotoh T, Hosokawa M, Piester D (2006) On the optimization of time transfer links for TAI. In: Proceedings of the European time and frequency forum. Braunschweig, Germany

Matsakis D, DeFraigne P, Banerjee P (2014) Review of precise time transfer. Radio Sci Bull (351):29–44

Time Reference, Calibration and Time Transfer Techniques for Satellite Altimetry

Elizabeth Laier English, Setnam Shemar, Kathryn Burrows, Conway Langham, Hannah Collingwood, and Peter Whibberley

Abstract

Synchronisation of a remote clock to a time reference can be challenging. Within the timing community these challenges have been addressed, and robust time transfer and calibration techniques have been developed offering differing levels of synchronization accuracy to the international time reference UTC, Coordinated Universal Time. These techniques can be applied to timing equipment at ground-based Cal/Val (Calibration/Validation) sites in Western Crete and elsewhere to achieve FRM (fiducial reference measurements) for altimetry, satisfying their requirement for SI (International System of units) traceability. Continuous monitoring of the remote sites is required to maintain traceability to the reference time, and a holdover clock may also be needed. This paper discusses how UTC or TAI (International Atomic Time) could be used as a time reference for timestamped measurements taken at Cal/Val sites, improving measurement uncertainty and linking fiducial reference measurements for satellite altimetry back to the SI unit of time: the second.

Keywords

Clocks · FRM · Satellite altimetry · Time reference · Time scale · Time transfer · UTC

1 Introduction

One of the objectives of the ESA (European Space Agency) workshop FRM4ALT was to establish how any international group working on satellite altimetry could attain SI traceability of their measurements and improve uncertainties. Satellite altimetry is used to monitor changes in sea-level continuously with mm/year accuracy with respect to the centre of mass of the Earth. The aim of referencing these altimetry measurements to the SI second is to reduce the uncertainty in altimetry satellite orbits (Ablain et al. 2017). Ultimately this will improve the accuracy in monitoring of sea-level changes, which is crucial for understanding long-term climate change, understanding oceans and explaining weather patterns.

There are many factors contributing to the uncertainty in sea-surface Cal/Val measurements: Absolute coordinates of the reference Cal/Val site location come from GNSS with the associated hardware and processing uncertainties, and the ground stability of the site and Earth tides also contribute (Mitchum 2000). Uncertainties in the water level determination depend on the site location and condition, the tide gauge used, and the local reference surface. Geophysical parameters (Mertikas 2011) and atmospheric delays in altimetry signals are also a factor. This shows the need for a reliable reference; the concept of FRM was invented by ESA for accurate standardization of the Cal/Val uncertainties in satellite measurements of sea-level.

The long-term global mean sea-level rise over 1993–2014 amounts to +3.4 (±0.4) mm/year (Ablain et al. 2017). This

This work was presented at ESA workshop FRM4ALT on 23rd April 2018.

E. L. English (✉) · S. Shemar · K. Burrows · C. Langham · H. Collingwood · P. Whibberley
National Physical Laboratory, Teddington, UK
e-mail: elizabeth.laier.english@npl.co.uk

S. P. Mertikas, R. Pail (eds.), *Fiducial Reference Measurements for Altimetry*, International Association of Geodesy Symposia 150, https://doi.org/10.1007/1345_2019_55

rise is due to ocean warming through thermal expansion of sea water and land ice melting, both of which result from anthropogenic global warming. However, the uncertainty of components of the sea level budget equation are of the order of 1 mm/year (2σ) (Church et al. 2013). ESA would like to standardize the ground based measurements and monitoring data taken from the Cal/Val transponders in SI units in order to improve these uncertainties. At the ground based calibration site CRS1 in southwest Crete, two tide gauges (a pressure tide gauge and a radar sensor) and a permanent GNSS station have been operational continuously from March 2008 (Ablain et al. 2017). The GNSS receivers at this and other sites in Western Crete could be calibrated to provide traceability at the level of ± 1 μs to UTC using the same time transfer and calibration techniques used by the timing community. In this way, fiducial reference measurements for altimetry could be referenced to the SI second.

1.1 UTC as a Reference

Coordinated Universal Time (UTC) is the international time scale which forms the basis of civil timekeeping worldwide. It is computed by the BIPM (Bureau International des Poids et Mesures) located in Paris from a weighted ensemble of around 500 atomic clocks in approximately 70 timing institutes around the world (Panfilo 2016). UTC is a time scale with the stability of atomic clocks but with occasional (∼yearly) 1-second adjustments known as leap seconds which maintain UTC within 0.9 s of UT1 (Earth rotation time). UTC exists only on paper and is published monthly by the BIPM in *Circular T* around the 10th of the month containing data for the previous month, resulting in measurement latency between 10 and 45 days. A 7-day 'rapid' approximation called UTCr is available on a weekly basis. Local real-time access to UTC, and traceability (Matsakis et al. 2018) for both time and frequency is provided by the UTC(k) timing institutes; in the UK this is UTC(NPL).

The irregular application of leap seconds to UTC, typically occurring on either 30 June or 31 Dec, may cause software issues. The IERS (International Earth Rotation and Reference Systems Service) decides 6 months in advance if a leap second will be required or not, and publishes this announcement in the "IERS Bulletin C". An alternative time scale called TAI (International Atomic Time) is a weighted average of atomic clocks calculated by the BIPM, which could be used as a reference instead (although TAI is not recommended as a reference by the BIPM as it is not formally disseminated). A formal definition of TAI and UTC adopted by the CGPM in November 2019 specifies the relativistic rate shift of a clock at the Earth's surface at a constant gravity potential, relative to a clock in zero gravity (Resolutions adopted by the BIPM at the 26th CGPM 2018).

A GNSS receiver can provide users with UTC and TAI, along with access to local GNSS system time such as GPS, which is operated by the US Naval Observatory (USNO) and steered to UTC(USNO) without including leap seconds. Correct installation and calibration of a GNSS receiver is essential if it is to provide an accurate time reference.

2 UTC(NPL) Time Transfer

UTC(NPL) is a typical example of a maser-based UTC(k) time scale. Between UTC(k) timing institutes, time transfer techniques with an uncertainty of a few ns are used; one is based on reception of GPS signals, and the other is TWSTFT (Two-Way Satellite Time and Frequency Transfer) via geostationary satellite. To disseminate time to a wide number of users, NTP (Network Time Protocol) can achieve clock synchronisation over packet-switched data networks. NTP is free to users, accessible over the internet and provides users with tens of ms level uncertainty to UTC. The UK's radio time signal, MSF, is also widely available: a time code is disseminated by a terrestrial radio transmitter connected to a time reference and is monitored by NPL. The uncertainty to UTC is a few ms.

3 Calibration Techniques and Uncertainties

Calibration of timing equipment is required for validation and consistency of measurements and can be carried out by an accredited laboratory. Delay differences inside the time transfer equipment must be measured to the required level of uncertainty. During calibration it is required to compare the remote clock inside the timing equipment with a UTC(k) reference. Methods are:

1. Bring the remote endpoint equipment to the reference clock for calibration; Endpoint timing equipment, e.g. GPS antenna and receiver can be shipped to a UTC(k) laboratory for calibration of the complete system. Measurements are usually obtained over 10 days to account for diurnal effects. A statistical analysis will determine the time offset to UTC(k) and measurement uncertainties. The calibration uncertainty is typically 5 ns (2σ).
2. Transport a third clock from the reference to the remote endpoint clock;

   A portable caesium clock can be transported from a UTC(k) laboratory to the endpoint timing equipment. The caesium clock is monitored against the UTC(k) reference before transport, and again on its return. The duration of the measurement at the endpoint location can range from 1 h to 10 days. A statistical analysis will determine the time offset to UTC(k) and measurement uncertainties. The calibration uncertainty is typically 2.5 ns (2σ) although this will depend on the caesium clock drift calculated over the duration of the calibration.

3. Use GPS for absolute or relative calibration at the remote endpoint clock; Absolute calibration requires a GNSS simulator operated in an anechoic chamber, which is not easily available. This method offers sub-ns uncertainty, although hardware delays are not fully determined (Görres et al. 2006). The relative calibration method is more commonly used, which compares GPS measurements collected by the endpoint receiving chain and a travelling reference receiver chain. The calibration uncertainty is of the order of 1 ns. This method is used for calibration of UTC(*k*) timing laboratory equipment; a travelling calibrated GPS receiver is used to calibrate UTC(*k*) labs' GPS receivers. This method of relative calibration is also applied to TWSTFT, using a travelling Earth station.

4 Monitoring and Holdover

Continuous monitoring of endpoint equipment is required for fault detection e.g. loss of reference signal. Time transfer or time dissemination techniques can be used for monitoring depending on the level of uncertainty to UTC required, for example: GPS Common View/All-in-View (where a satellite signal is received at different locations simultaneously) uncertainty is a few ns to UTC; NTP via the internet provides tens of ms and radio time signals (e.g. MSF) provide a few ms to UTC. If optical fibre is available to link the remote sites to a reference, higher accuracy techniques exist e.g. PTP (Precision Time Protocol) could provide hundreds of ns uncertainty to UTC.

The monitoring system can only show that a discrete time step has occurred in the data. To maintain accurate time if reference synchronisation has been lost, a holdover clock may be required. Many atomic clocks are commercially available with a wide range of performance, size and cost. At the high end are hydrogen masers, followed by the more widely used caesium clock. Rubidium oscillators are typically found in GPS disciplined oscillators. Chip scale atomic clocks (CSACs) offer a low size, weight and power option, although with lower stability. Quartz is the most commonly used low stability clock.

5 Relativistic Effects

The frequency of a clock on-board a satellite will be affected by relativistic effects relative to a clock on the ground (Nelson 2011). For example, with respect to a clock on the geoid, a clock on a satellite in a circular orbit of altitude 1,000 km (for altimetry satellites), the rate is slower by approximately 18 μs/day. The contribution from the difference in gravitational potential to this net rate is 8 μs/day faster, and the difference in velocity is 26 μs/day slower.

The effect of the Earth's gravity potential on the frequency of a clock is 10^{-13} per km of altitude close to the surface, and the uncertainty of the gravity potential on the geoid equates to a frequency uncertainty of approximately 1×10^{-17} s/s (Riehle 2006). TCG (Geocentric Coordinate Time) is equivalent to the proper time at the centre of the Earth and is not influenced by gravitational effects. TCG is a coordinate time not a time scale; there are no practical realizations of TCG. A clock ticking at the rate of TCG ticks faster than a clock ticking at the scale of UTC by a factor of $6.969290134 \times 10^{-10}$ or 60.2 μs/day (Resolutions adopted by the BIPM at the 26th CGPM 2018; Nelson 2011).

6 Summary

This paper presents an overview of how UTC or TAI time scales could be used as a time reference for satellite altimetry, reducing Cal/Val site uncertainties. Many of the calibration and time transfer methods used by the timing community could be applied to FRM for altimetry, providing a reference to the SI second.

References

Ablain M, Legeais JF, Prandi P, Marcos M, Fenoglio-Marc L, Dieng HB, Benveniste J, Cazenave A (2017) Satellite altimetry-based sea level at global and regional scales. Surv Geophys 38:7–31. https://doi.org/10.1007/s10712-016-9389-8

Church JA, Clark PU, Cazenave A, Gregory JM, Jevrejeva S, Levermann A, Merrifield MA, Milne GA, Nerem RS, Nunn PD, Payne AJ, Pfeffer WT, Stammer D, Unnikrishnan AS (2013) Sea level change. In: Climate change 2013: the physical science basis. Cambridge University Press, Cambridge, pp 1137–1216. http://www.climatechange2013.org/images/report/WG1AR5_Chapter13_FINAL.pdf

Görres B, Campbell J, Becker M, Siemes M (2006) Absolute calibration of GPS antennas: laboratory results and comparison with field and robot techniques. GPS Solutions 10(2):136–145. https://doi.org/10.1007/s10291-005-0015-3

Matsakis D, Levine J, Lombardi M (2018) Metrological and legal traceability of time signals. In: Proceedings of the 49th Annual PTTI Meeting, Reston, Virginia, January, pp 59–71. https://www.ion.org/publications/abstract.cfm?articleID=15605

Mertikas S (2011) Geodesy, ground positioning and leveling, encyclopedia of solid earth geophysics. Encyclopedia of earth sciences series. Springer, Dordrecht. https://doi.org/10.1007/978-90-481-8702-7_178

Mitchum G (2000) An improved calibration of satellite altimetric heights using tide gauge sea levels with adjustment for land motion. Marine Geodesy 23(3):145–166. https://doi.org/10.1080/01490410050128591

Nelson RA (2011) Relativistic time transfer in the vicinity of the Earth and in the solar system. Metrologia 48:S171–S180. http://stacks.iop.org/Met/48/S171

Panfilo G (2016) The coordinated universal time. IEEE Instrum Meas Mag 19:28–33. https://doi.org/10.1109/MIM.2016.7477951

Resolutions adopted by the BIPM at the 26th CGPM, Nov 2018. https://www.bipm.org/utils/common/pdf/CGPM-2018/26th-CGPM-Resolutions.pdf

Riehle F (2006) Frequency standards: basics and applications. Wiley, Hoboken, pp 460–464. https://doi.org/10.1002/3527605991. ISBN: 978-3-527-60595-8

Calibrating CryoSat-2 and Sentinel-3A Sea Surface Heights Along the German Coast

Luciana Fenoglio, Salvatore Dinardo, Christopher Buchhaupt, Bernd Uebbing, Remko Scharroo, Jüergen Kusche, Matthias Becker, and Jérôme Benveniste

Abstract

By convention the absolute bias in sea surface height (SSH) is the difference between the altimeter and the in-situ reference SSH heights above the Earth ellipsoid. Both the absolute and the relative bias of the CryoSat-2 and Sentinel-3A missions are derived in this study at four stations along the German coasts.

Firstly, the coastal data processed in Delay Doppler altimeter (DDA) mode, also called SAR mode (SARM), are shown to be less noisy than data in pseudo-low resolution mode (PLRM), which is comparable to the conventional low-resolution mode (LRM). The best agreement with in-situ data is reached by the SARM data retracked with the SAMOSA+ coastal retracker (hereafter SAR/SAMOSA+) from the ESA GPOD SARvatore service.

Secondly the absolute bias and its standard deviation are computed for each mission and product type. The mean mission absolute bias depends on location and altimeter product. Both the absolute and relative biases are small and the standard deviation is smaller than 4 cm and larger than the bias. Departures between absolute biases evaluated at different stations are possibly related to geoid inaccuracy in the coastal zone. Finally, the smaller standard deviation of the bias time series confirms that SAR altimetry is more accurate than PLRM, the minimum standard deviation is 2 cm.

Keywords

Calibration · Coastal altimetry · CryoSat-2 · Delay-Doppler · SAR altimetry · Sea surface height · Sentinel-3A · Tide gauge

1 Introduction

The calibration of the altimeter data range insures the continuity of the altimetric missions, thus allowing to derive regional sea level variations with accuracy better than 1 mm/year, which is the actual challenge (Ablain et al. 2017). The absolute bias of each mission and its change are monitored at dedicated Calibration/Validation (cal/val) sites by direct comparison of the altimetric data with in-situ data (Christensen et al. 1994; Haines et al. 2003; Watson et al. 2011; Bonnefond et al. 2018). Specific calibrations scenarios, as in Crétaux et al. (2013) who perform GPS surveys from a boat cruising along the satellite tracks, and relative calibration approaches as in Cancet et al. (2013), are also used. The relative biases between the missions is

L. Fenoglio (✉) · C. Buchhaupt · B. Uebbing · J. Kusche
Institute of Geodesy and Geoinformation, University of Bonn, Bonn, Germany
e-mail: fenoglio@geod.uni-bonn.de

S. Dinardo
HeSpace, Darmstadt, Germany

R. Scharroo
EUMETSAT, Darmstadt, Germany

M. Becker
Institute of Geodesy, Technical University Darmstadt, Darmstadt, Germany

J. Benveniste
ESA/ESRIN, Frascati, Italy

S. P. Mertikas, R. Pail (eds.), *Fiducial Reference Measurements for Altimetry*,
International Association of Geodesy Symposia 150, https://doi.org/10.1007/1345_2019_73

the difference of the absolute calibration biases and can be determined independently by crossover analysis.

Following Shum et al. (2003), a large number of cal/val sites is necessary to have a robust assessment of the radar altimeter data over different types of water surfaces. In this study, we determine the absolute bias at four tide gauge stations co-located with GPS stations along the German coasts of the North Sea and Baltic Sea. The Sentinel-3A and CryoSat-2 missions include altimeters in synthetic aperture radar mode (SARM). The advantages of SARM data with respect to PLRM, also called Reduced SAR (RDSAR), were reported for this region by Fenoglio-Marc et al. (2015) and Dinardo et al. (2018) and globally by several authors (e.g. Boy et al. 2017).

Section 2 describes briefly data and methodology. In Sect. 3 after an assessment of coastal data quality, bias and its variation are computed for both SARM and PLRM. The results are discussed and conclusions are drawn in Sect. 4.

2 Methodology and Data

Sea Surface ellipsoidal heights are measured by two approaches: (1) directly from satellite altimeter and (2) from tide gauge stations colocated with a GPS permanent station. Knowledge of the vertical distance between the zero of the tide gauge station and the GPS benchmark allows in the second approach to transform the tide gauge observations in ellipsoidal heights of water level. All heights are transformed beforehand to the mean tide system, which is the system used in satellite altimetry. The different location of the altimeter and tide gauge measurements is accounted for by correcting the ellipsoidal heights for the geoid height difference between the two locations.

Along the German coast of North and Baltic Sea (Fig. 1), quite a number of tides gauge stations are colocated with GPS stations. We consider four of them, namely Helgoland (HELG), Sassnitz (SASS), Warnemuende (WARN), Kiel Holtenau (TGKI). Further on, we consider the Lighthouse Kiel station (LHKI), not colocated to GPS, which has a very good agreement between altimeter and in-situ variability. Few peculiarities of the region help to interpret the results. The geoid gradients are larger in the Baltic than in the North Sea (see Figs. 1 and 4). The ocean tides are smaller in the Baltic than in North Sea and affect the bias mainly in the coastal zone, while in open sea they are well predicted by tidal models.

The Sentinel-3A and CryoSat-2 satellites carry altimeters in SARM, which differ from conventional pulse-bandwidth limited altimeters. The altimeter data cover 91 months of CryoSat-2 (from October 2010 to May 2018, hereafter called periodA) and 18 months of Sentinel-3A (from June 2016 to December 2017, hereafter called periodB).

The three types of CryoSat-2 products used are a SAR and two PLRM products. The GPOD SAR product, hereafter SAR/SAMOSA+, is processed with the options adopted in Dinardo et al. (2018), i.e. Hamming weighting window on the burst data prior to the azimuth FFT, zero-padding prior to the range FFT and doubling of the extension for the radar range swath. The two PLRM products are in-house products using the STAR (Roscher et al. 2017) and TALES (Dinardo et al. 2018) retrackers. They are called PLRM/STAR and PLRM/TALES hereafter.

The three types of Sentinel-3A products used are two SAR and one PLRM products. The PLRM/MARINE product uses an open ocean retracker. The SAR official Copernicus Marine product uses the SAMOSA2 open ocean retracker, while the second SAR product is SAR/SAMOSA+ as for CryoSat-2.

Homogeneous corrections for the wet atmospheric (GPD+, Fernandes and Lázaro 2016) and the ocean tide (OSU TPXO8-ATLAS model, Egbert and Erofeeva 2002) corrections are applied to all altimeter datasets. For the instantaneous comparison with the tide gauges the methodology defined in Fenoglio-Marc et al. (2015) is followed with the ocean tide applied and the dynamic atmospheric correction (DAC) not applied. For the noise investigation all corrections are applied to the altimeter data. A constant value of +0.025 m has been added to the CryoSat-2 SAR/SAMOSA+ sea surface height (SSH) and sea level anomaly (SLA) for consistency with the standard CryoSat-2 processing, which uses a static bias of −0.6730 m instead of the −0.6980 m used in the SARvatore GPOD (Fenoglio-Marc et al. 2015). The same value is used by Bonnefond (OSTST 2018, https://www.aviso.altimetry.fr).

Weekly ellipsoidal coordinates of the GPS benchmark of the four stations have been made available from BKG in the International Reference Frame System (ITRF2008) (Rülke et al. 2008). This enables the monitoring of time-dependent vertical land motion (VLM). The tide gauge LTKI instead is not colocated with a GPS, but the standard deviation of the differences between altimeter and in-situ data anomalies is minimum.

Firstly, the performance of the altimeter data is investigated in terms of noise level as function of the distance to coast using two different approaches to describe the noise. In one approach the absolute differences of consecutive 20 Hz SLAs are computed and averaged as function of the distance to coast. In the other approach the standard deviations of altimetric and model SLAs are compared. The BSHcmod model is from the German Federal Maritime and Hydrographic Agency (BSH) and was used in Fenoglio-Marc et al. (2015), Dinardo et al. (2018) and Schall et al. (2016) as well.

Secondly, the accuracy and bias of the altimeter sea surface height measurements is investigated by comparing 1 Hz altimeter and in-situ SSHs above the reference ellipsoid.

For Cryosat-2 the two intervals periodA and periodB are considered. The metrics is given by mean, standard deviation of differences and correlation between altimeter and tide gauge data.

3 Results

In agreement with Dinardo et al. (2018), we found that data processed in SAR mode outperform the PLRM data in the coastal zone. This is shown here by the evolution as function of the distance to coast of the two parameters: (1) the noise of the along-track data and (2) the standard deviation of SLAs.

Firstly, Fig. 2 gives for Sentinel-3A the scatterplot of the noise of the SLAs in 1-km bins of distance from the coast for SAR (left) and PLRM (right). The median of the distribution is a good indicator of the level of noise. In SAR the median stays flat at 5 cm from the open ocean up to 2 km from the coast and is still relatively low (25 cm) at 1 km from the coast. The 25th and the 75th percentiles stay flat at 2 cm and 8 cm and rises at 1 and 4 km. The noise for PLRM is larger, with the median of the distribution staying flat at 11 cm from the open ocean up to 7 km and then increasing to 14 cm up to 2 km from the coast.

Then, the standard deviations of altimeter and model SLA are investigated against the distance to the coast. Figure 3 shows that the best agreement is obtained by the SAR/SAMOSA+ data and the second best agreement with the PLRM/STAR. The official marine data do not include a coastal retracker and perform worst near coast, the worst results correspond to the marine PLRM data.

Further on, time-series of differences between the altimeter and the in-situ sea surface heights above the ellipsoid are built for each missions, data product and time interval at each of the four colocated tide gauge stations. Instead, at the LTKI station the time-series of differences between the altimeter and the in-situ SLAs is built and finally only the de-meaned time-series is considered as in this case the altimeter bias cannot be estimated.

To build the time series we select the nearest altimeter data to the tide gauge which is located within a selected range of distances from the tide gauge station; the range of distances is defined by a minimum and maximum radius. The interval is 5–10 km in coastal zone and 10–20 km in open sea, but we enlarge this spatial constraint for the Sentinel-3A data depending on the availability of 1 Hz data in this interval of distances. The locations of the CryoSat-2 and Sentinel-3A points are shown for Helgoland and Sassnitz in Fig. 4. The 1 Hz Sentinel-3A data are all located near to each other on the same ground-track and their number equals the number of cycles considered. Instead, the 1 Hz CryoSat-2 data are on several ground tracks and spread in a larger area with their number increasingly linearly with the search radius. The CryoSat-2 points corresponding to the two intervals periodA and periodB are shown in Fig. 4 in green and purple respectively.

We estimate mean and standard deviation of the time-series, the mean is the absolute bias of the mission. Table 1 summarizes the SAR/SAMOSA+ absolute biases (Ca and Cb for CryoSat-2 over the two time intervals and S3 for Sentinel-3A) and the relative biases over the common interval (S3-Cb) and time-series standard deviation (StCa, StCb and StS3) over each interval. The Helgoland time-series over the complete interval are shown in Figs. 5 and 6, the Sassnitz time-series are in Figs. 7 and 8. We report here the values corresponding to the SAR/SAMOSA+, as this is the most accurate product. We observe that in Helgoland the CryoSat-2 mean biases are similar over the complete and the reduced interval. At other stations they differ, which suggests that a long time interval is preferable for the analysis. The CryoSat-2 bias relative to the Helgoland tide gauge is −31 cm over the Sentinel-3A time interval. The mean bias over the Sentinel-3A mission is −36 mm and the relative bias between the mis-

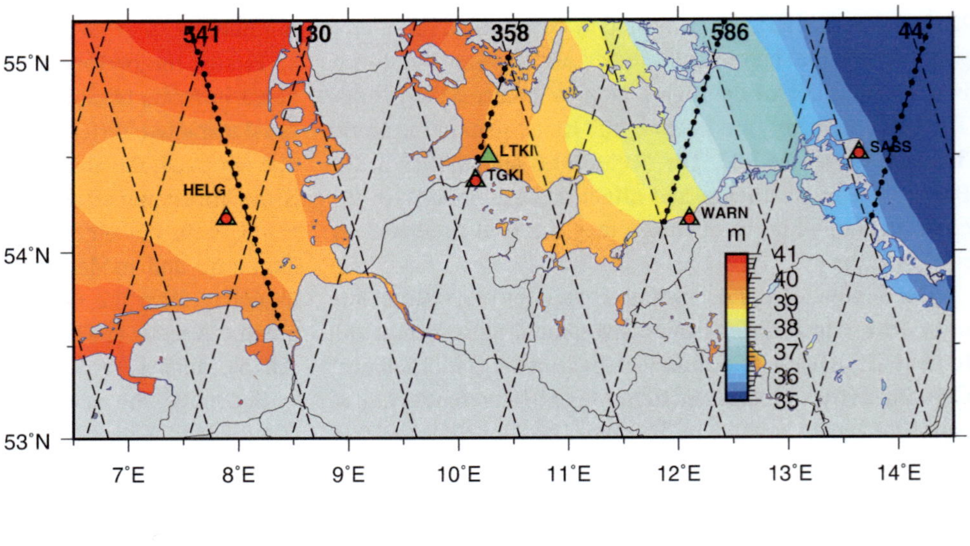

Fig. 1 Region of analysis with altimeter tracks and in-situ tide gauge (triangle), GPS (circle) stations. The German Combined QuasiGeoid 2016 (GCG2016) is in colored map

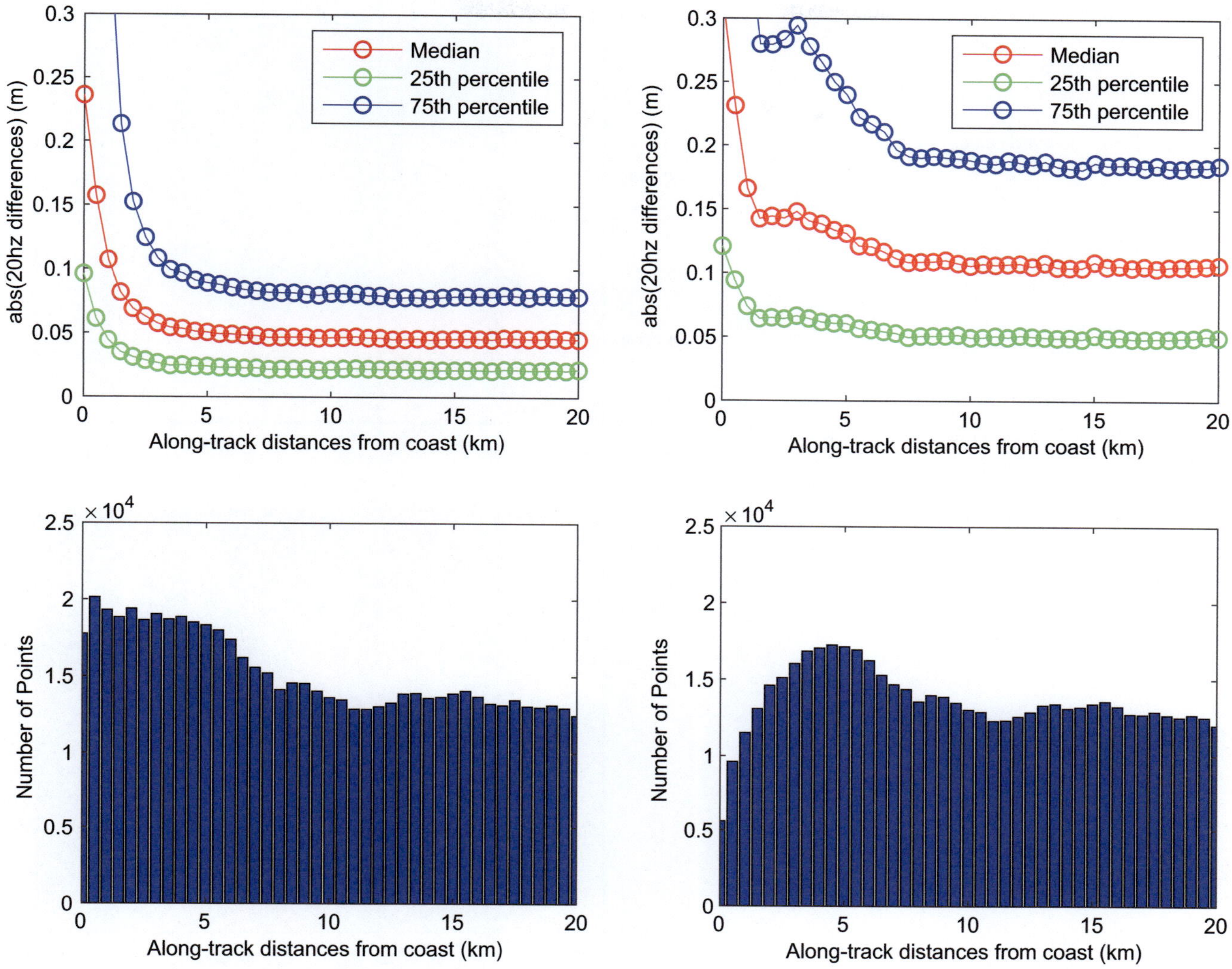

Fig. 2 Noise of 20-Hz Sentinel-3A SAR Marine (left) and PLRM (right) Marine altimeter SLA as function of distance to coast (top) and number of retained data (bottom)

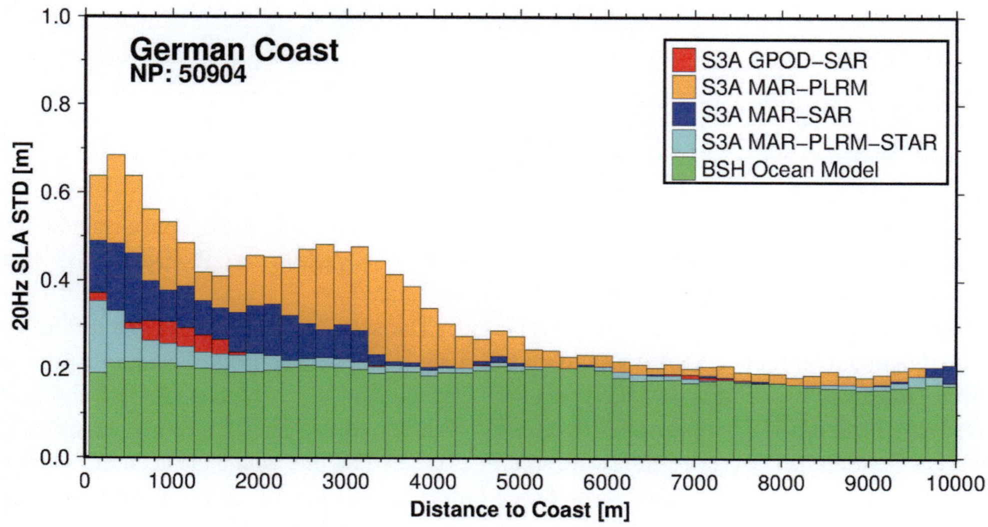

Fig. 3 Standard deviation of Sentinel-3A SLAs as function of distance to coast for two SAR and two PLRM products

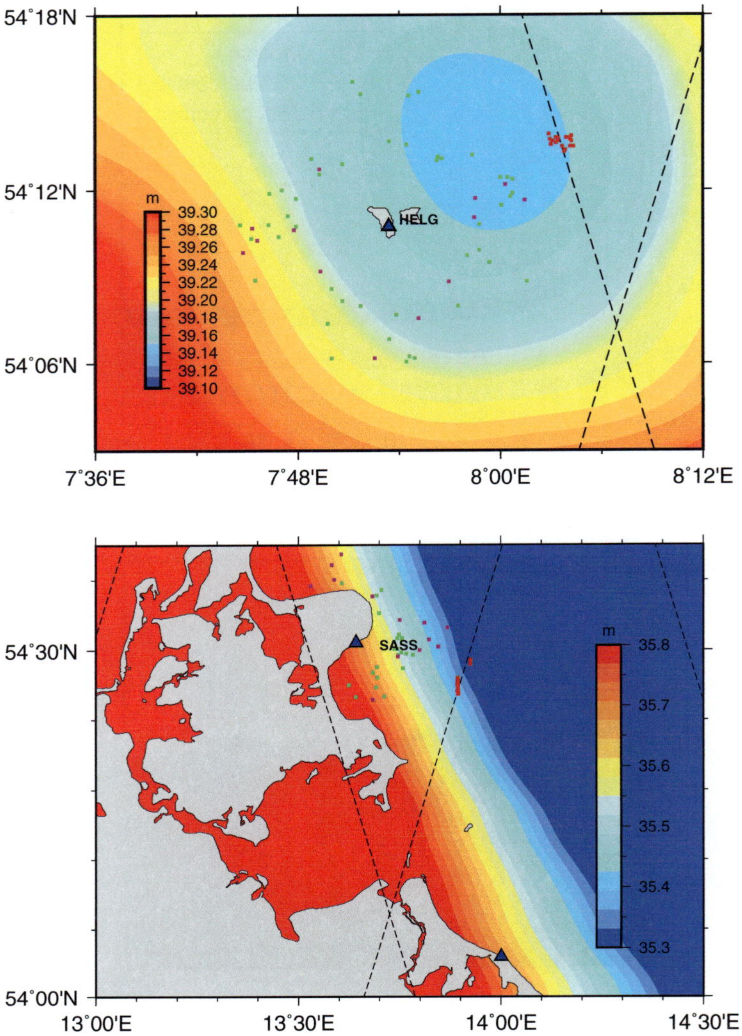

Fig. 4 CryoSat-2 and S-3 1 Hz data at distances 5–10 km from the tide gauge in Helgoland (above) and in Sassnitz (below). CryoSat-2 PeriodA in green, CryoSat-2 PerioB in purple and S-3 in red. The quasigeoid GCG2016 is in colored map

Table 1 For SAR/SAMOSA+ processing: Bias (S3-Cb) and standard deviation (mm) of altimeter and tide gauge differences for CryoSat-2 complete interval (Ca) and CryoSat-2 short interval (Cb) and Sentinel-3A

Station	Ca	Cb	S3	S3-Cb	StCa	StCb	StS3
HELG	−33	−31	−36	−5	40	33	65
SASS	0	24	+13	−11	35	59	36
WARN	−45	−2	−54	−52	40	38	38
TGKI	−28	−23	−3	20	42	26	33

sions is 5 mm. A similar conclusion is drawn from the results in Sassnitz where the CryoSat-2 bias is 24 mm over the Sentinel-3A time interval and the mean bias of Sentinel-3A is higher than in Helgoland, finally leading to a satellite relative bias of 1.1 cm. In Helgoland both the SAR/SAMOSA+ and PLRM/STAR time-series have standard deviation smaller than 4 cm. Instead, the PLRM/TALES time-series is highly variable with outliers. For Sentinel-3A the residuals are small for all the modes and the standard deviation is between 5 and 7 cm. Also in Sassnitz the SAR/SAMOSA+ and PLRM/STAR time-series have standard deviation smaller than 4 cm, while the PLRM/TALES time-series is more variable.

The standard deviation of the difference with the in-situ data is an estimate of the accuracy of the altimeter measurements. The mean of the standard deviations in Table 1 is 40 mm and indicates a good accuracy of the sea level data. At the LTKI tide gauge the accuracy is higher, with standard deviation 21 mm using altimeter data in open sea (10–20 km from coast) (Fig. 9).

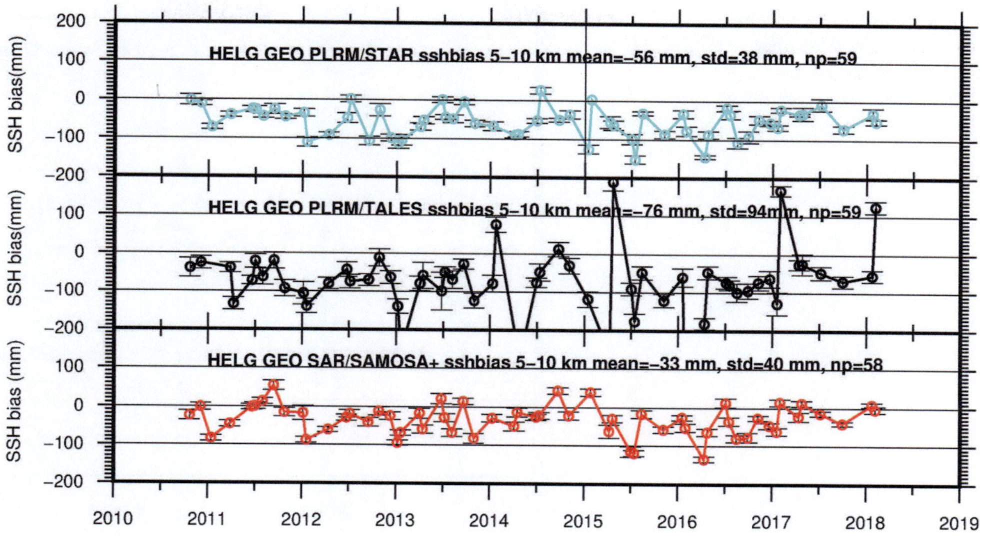

Fig. 5 CryoSat-2 bias at Helgoland for SAR/SAMOSA+, PLRM/TALES and PLRM/STAR data

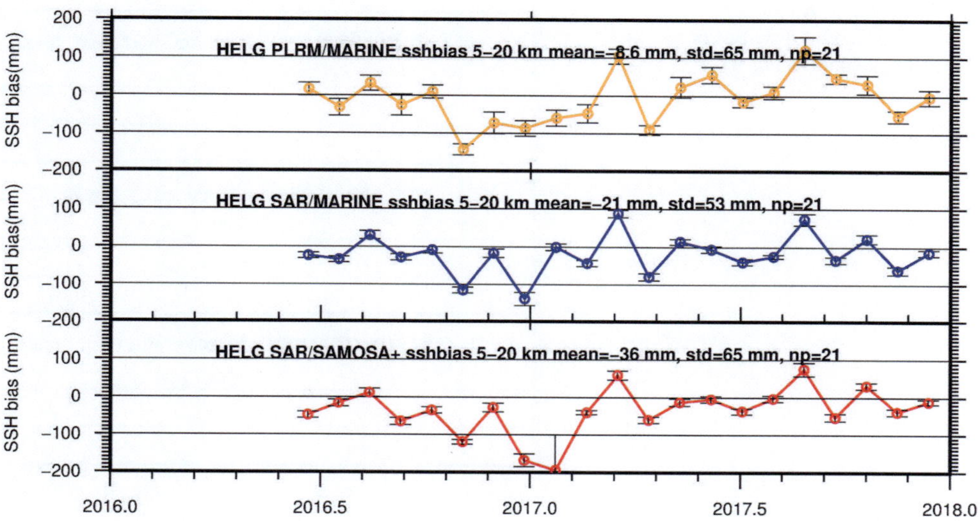

Fig. 6 Sentinel-3A bias at Helgoland for SAR/Marine, PLRM/Marine and SAR/SAMOSA+, colours as Fig. 3

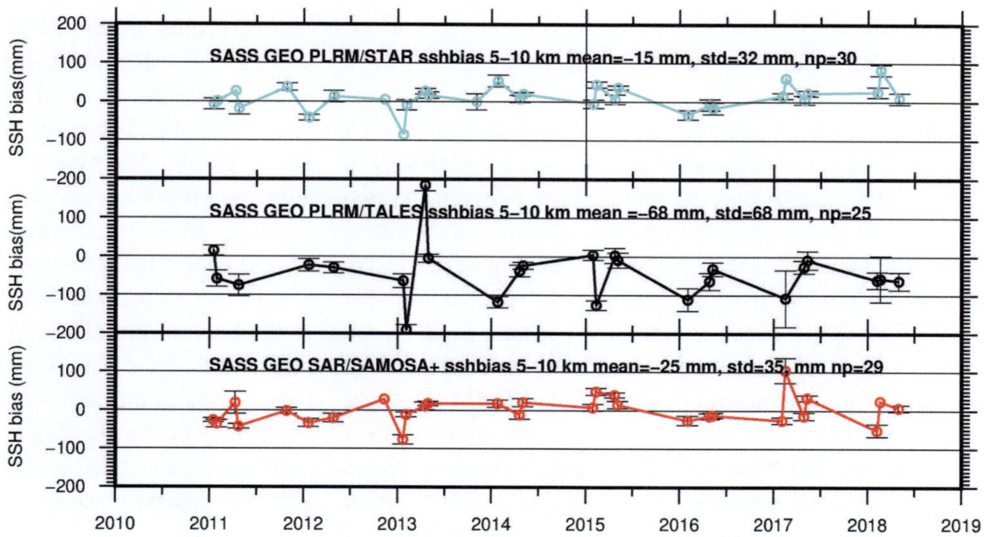

Fig. 7 CryoSat-2 bias in Sassnitz for SAR/SAMOSA+, PLRM/TALES and PLRM/STAR

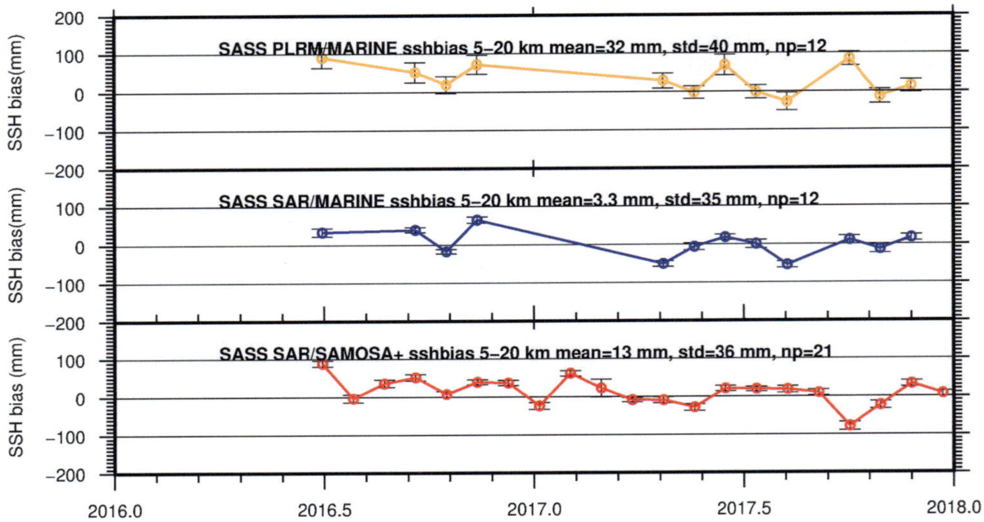

Fig. 8 Sentinel-3A bias in Sassnitz for SAR/Marine, PLRM/Marine and SAR/SAMOSA+, colours as Fig. 3

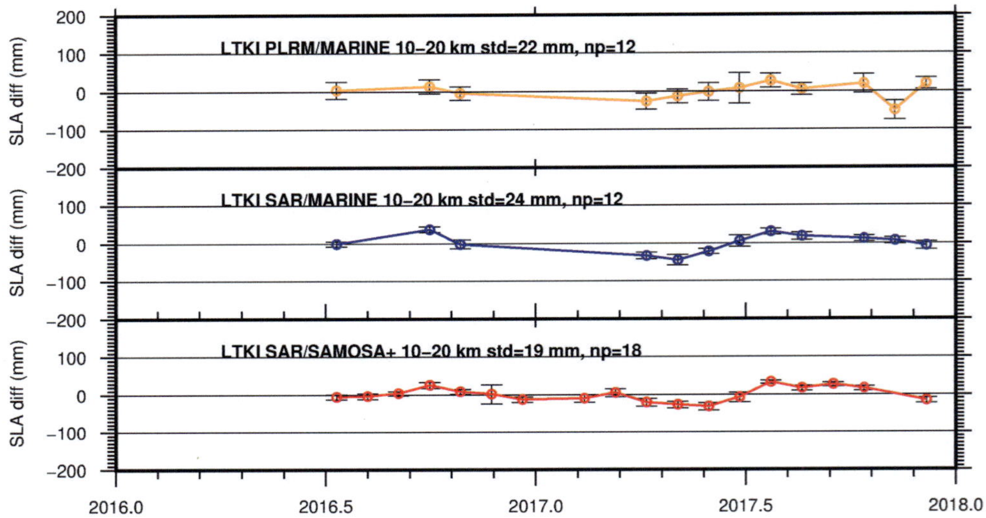

Fig. 9 Sentinel-3A bias at Lighthouse Kiel for distances of 10–20 km, colours as Fig. 3

4 Discussion and Conclusion

The absolute biases of the CryoSat-2 and of the Sentinel-3A altimeter measurements have been evaluated in Germany at four coastal stations and the relative bias between the two missions was derived as difference of the absolute biases. The absolute biases differ by 1–5 cm depending on tide gauge station and product type, the same for the relative bias. For SAR/SAMOSA+, which is the most accurate product, the relative bias is smaller or equal to 10 mm in Helgoland and Sassnitz, in agreement with Bonnefond et al. (2018). A larger relative bias is found at the other two stations, with 20 mm in Kiel Holtenau and 50 mm in Warnemuende. Absolute biases for CryoSat-2 and Sentinel-3A are −31 mm and −36 mm in Helgoland and 13 mm and 0 mm in Sassnitz respectively.

Finally, the standard deviation of the bias time series has been analysed pro mission and product type. The standard deviation of the estimated bias at a given station is between 2 and 4 cm. The higher stability of the SAR SSH bias compared to the PLRM SSH bias, confirms the higher accuracy of SAR over PLRM altimetry, see Figs. 5, 6, 7, and 8. The minimum standard deviation is 19 mm at Lighthouse Kiel. Moreover, compared to PLRM data which are generally unusable closer than 5 km from the coast, SAR mode altimeters provide reliable measurements also at less than five Kilometers from the coast.

The study shows that the German network of stations is suitable for cal/val analysis, however the standard deviation of the bias is high and need to be reduced for a significant assessment of the bias. The detected differences in absolute bias derived at different stations are possibly caused by the inaccuracy of the geoid in coastal zone and are expected to reduce with dedicated analysis. In further studies the 20 Hz altimeter data will be used.

Acknowledgements The authors acknowledge EUMETSAT, the European Space Agency (ESA) and the ESA/ESRIN G-POD team for the CryoSat-2 and Sentinel-3A data products. In-situ data are kindly provided by the German Waterway and Shipping Administration (WSV) and by the German Federal Institute of Hydrology (BfG).

Local geoid GCG2016 and GPS coordinates were make available by the Federal Agency for Cartography and Geodesy (BKG). We acknowledge the German Federal Maritime and Hydrographic Agency (BSH), the Deutscher Wetterdienst (DWD), the European Centre for Medium-range Weather Forecasts (ECMWF) for providing model data. Support to the first author was partly provided by ESA in the frame of the Sea Level Climate Change Initiative (SLCCI, ESA Contract N.4000109872/14/I-NB), the GOCE++ DYnamic COastal Topography and tide gauge unification (GOCE/Dycot, ESA Contract N.4000114331/15/NL-FF) and the SAR Altimetry Coastal and Open Ocean Performance (SCOOP, ESA Contract 4000115385/15/I-BG) projects. We are also grateful to the editor and to the two anonymous reviewers who helped to improve the manuscript.

References

Ablain M, Legeais JF, Prandi P, Marcos M, Fenoglio-Marc L, Dieng HB, Benveniste J, Cazenave A (2017) Satellite altimetry-based sea level at global and regional scales. Surv Geophys 38(1):7–31. https://doi.org/10.1007/s10712-016-9389-8

Bonnefond P, Laurain O, Exertier P, Boy F, Guinle T, Picot N, Labroue S, Raynal M, Donlon C, Féménias P, Parrinello T, Dinardo S (2018) Calibrating the SAR SSH of Sentinel-3A and CryoSat-2 over the Corsica facilities. Remote Sens 10(2):92. https://doi.org/10.3390/rs10010092

Boy F, Desjonqueres JD, Picot N, Moreau T, Raynal M (2017) CryoSat-2 SAR-Mode over oceans: processing methods, global assessment, and benefits. IEEE Trans Geosci Remote Sens 55(1):148–158. https://doi.org/10.1109/TGRS.2016.2601958

Cancet M, Bijac S, Chimot J, Bonnefond P, Jeansou E, Laurain O, Lyard F, Bronner E, Féménias P (2013) Regional in situ validation of satellite altimeters: calibration and cross-calibration results at the Corsican sites. Adv Space Res 51(8):1400–1417. https://doi.org/10.1016/j.asr.2012.06.017

Christensen EJ, Haines BJ, Keihm SJ, Morris CS, Norman RA, Purcell GH, Williams BG, Wilson BD, Born GH, Parke ME, Gill SK, Shum CK, Tapley BD, Kolenkiewicz R, Nerem RS (1994) Calibration of TOPEX/POSEIDON at platform harvest. J Geophys Res 99(C12):24465. https://doi.org/10.1029/94JC01641

Crétaux JF, Bergé-Nguyen M, Calmant S, Romanovski V, Meyssignac B, Perosanz F, Tashbaeva S, Arsen A, Fund F, Martignago N, Bonnefond P, Laurain O, Morrow R, Maisongrande P (2013) Calibration of Envisat radar altimeter over Lake Issykkul. Adv Space Res 51(8):1523–1541. https://doi.org/10.1016/j.asr.2012.06.039

Dinardo S, Fenoglio-Marc L, Buchhaupt C, Becker M, Scharroo R, Joana Fernandes M, Benveniste J (2018) Coastal SAR and PLRM altimetry in German Bight and West Baltic Sea. Adv Space Res 62(6):1371–1404. https://doi.org/10.1016/j.asr.2017.12.018

Egbert GD, Erofeeva SY (2002) Efficient inverse modeling of barotropic ocean tides. J Atmos Oceanic Tech 19(2):183–204. https://doi.org/10.1175/1520-0426(2002)0190183:EIMOBO2.0.CO;2

Fenoglio-Marc L, Dinardo S, Scharroo R, Roland A, Dutour Sikiric M, Lucas B, Becker M, Benveniste J, Weiss R (2015) The German Bight: a validation of CryoSat-2 altimeter data in SAR mode. Adv Space Res 55(11):2641–2656. https://doi.org/10.1016/j.asr.2015.02.014

Fernandes M, Lázaro C (2016) GPD+ Wet tropospheric corrections for CryoSat-2 and GFO altimetry missions. Remote Sens 8(10):851. https://doi.org/10.3390/rs8100851

Haines BJ, Dong D, Born GH, Gill SK (2003) The harvest experiment: monitoring Jason-1 and TOPEX/POSEIDON from a California offshore platform special issue: Jason-1 calibration/validation. Mar Geod 26(3–4):239–259. https://doi.org/10.1080/714044520

Roscher R, Uebbing B, Kusche J (2017) STAR: Spatio-temporal altimeter waveform retracking using sparse representation and conditional random fields. Remote Sens Environ 201:148–164. https://doi.org/10.1016/j.rse.2017.07.024

Rülke A, Dietrich R, Fritsche M, Rothacher M, Steigenberger P (2008) Realisation of the terrestrial reference system by a reprocessed global GPS network: realization of the TRS. J Geophys Res Solid Earth 113(B8). https://doi.org/10.1029/2007JB005231

Schall J, Löcher A, Kusche J, Rietbroek R, Sudau A (2016) Consistency of geoid models, radar altimetry, and hydrodynamic modelling in the North Sea. Mar Geod 39(3–4):223–237. https://doi.org/10.1080/01490419.2016.1152334

Shum C, Yi Y, Cheng K, Kuo C, Braun A, Calmant S, Chambers D (2003) Calibration of JASON-1 altimeter over Lake Erie special issue: Jason-1 calibration/validation. Mar Geod 26(3–4):335–354. https://doi.org/10.1080/714044525

Watson C, White N, Church J, Burgette R, Tregoning P, Coleman R (2011) Absolute calibration in Bass Strait, Australia: TOPEX, Jason-1 and OSTM/Jason-2. Mar Geod 34(3–4):242–260. https://doi.org/10.1080/01490419.2011.584834

Performance of Sentinel-3A SAR Altimetry Retrackers: The SAMOSA Coastal Sea Surface Heights for the Baltic Sea

Elzbieta Birgiel, Artu Ellmann, and Nicole Delpeche-Ellmann

Abstract

Performance of the Sentinel-3A SRAL altimeter data is evaluated in the coastal waters of the Gulf of Finland, Baltic Sea. Sea Surface Heights (SSH) were computed along three ascending passes nearby three tide gauge (TG) sites in Estonia. These SSH were compared to a high-resolution marine geoid model in conjunction with a TG corrected hydrodynamic model (HDM). The SAMOSA2 and SAMOSA+ retracker retrieved SSH values were inter-compared as well. The quality assessment yielded the root mean square error of 115 and 99 mm for the SAMOSA2 and SAMOSA+ retrieved Sentinel-3A SSH data, respectively. The near-zero mean of discrepancies shows that there is no significant systematic bias between the geodetic infrastructure and Sentinel-3A SSH data.

Keywords

Coastal altimetry · Gulf of Finland · Hydrodynamic model · Marine geoid · SAMOSA · Sentinel-3A · Tide gauges

1 Introduction

The Baltic Sea located in Northern Europe, is a semi-enclosed narrow sea that is characteristically known for many (sub-)basins, numerous archipelagos and its relatively shallow depth (in average 54 m). The marine traffic to the Baltic Sea countries is very intense, therefore the accurate Sea Surface Heights (SSH) and the marine geoid model become of utmost relevance, as these can be extremely useful for applications, such as marine engineering and ocean sciences. Classical computations of SSH have utilised tide gauges (TG) that are referenced to a particular vertical datum (see e.g. Woodworth et al. 2015; Liibusk et al. 2013). This method, however is limited by several factors, with the most relevant being spatial scales, as TG-s are land bounded. Satellite altimetry (SA), however, is not land bounded and instead provides offshore coverage. Numerous corrections need to be applied to improve the accuracy of satellite derived SSH. Also, accurate computations of SSH within the coastal area have remained a challenge, mainly due to land and calm water interference, uncertainty in the modelling of high-frequency tidal and atmospheric forcing etc. To overcome some of these obstacles several SA retrackers have been developed to improve the SSH accuracy especially in the coastal areas (Passaro et al. 2014; Bonnefond et al. 2018). Nonetheless, for validation of SA data the TG records in conjunction with a national datum related geoid model are usually utilised (e.g. Varbla et al. 2017). The nations around the Baltic Sea have modernized their geodetic infrastructure and established a dense network of TG stations. Exceptionally, it is also a region with accurate high resolution marine geoid models. This makes the Baltic Sea a suitable test site for validation and calibration of altimetry data, a sort of "marine laboratory". In addition, regionally adapted hydrodynamic

E. Birgiel (✉) · A. Ellmann
Department of Civil Engineering and Architecture, Tallinn University of Technology (TUT), Tallinn, Estonia
e-mail: elzbieta.birgiel@ttu.ee; artu.ellmann@ttu.ee

N. Delpeche-Ellmann
Department of Cybernetics, Tallinn University of Technology, Tallinn, Estonia
e-mail: nicole.delpeche@ttu.ee

S. P. Mertikas, R. Pail (eds.), *Fiducial Reference Measurements for Altimetry*,
International Association of Geodesy Symposia 150, https://doi.org/10.1007/1345_2019_59

models have the potential in complementing and contributing to offshore validations. In this study Sentinel-3A SAR Radar Altimeter (SRAL) products originating from the European Space Agency (ESA) developed SAMOSA2 (SAR Altimetry MOde Studies and Applications, see details in Dinardo et al. 2015) and SAMOSA+ retrackers are examined. The latter is an upgrade of the SAMOSA2 retracker, that may enable better usability over inland waters, sea ice, and the marine coastal zones (such comparisons can be found in Bonnefond et al. 2018). It is important to access the performance of the SRAL SSH and to determine its reliability for the Baltic Sea, especially within the complex coastal zones. Therefore, the SRAL derived SSH using two specific retrackers along the satellite tracks are compared with in-situ records via a high-resolution geoid model. To our present knowledge, this is one of the few studies (e.g. Dinardo et al. 2017) that attempts to assess the actual performance of the retracker derived SSH in the near-coast zones of the Baltic Sea, by a high-resolution geoid model. The methodology employed requires post processing and analysis of altimetry, TG and model data. For offshore verifications, the TG records are complemented with a regionally adapted hydrodynamic model (HDM). All terrestrial data sets are referenced to the same datum of the SA derived SSH (i.e. with respect to the reference ellipsoid) using the local geoid model. An assessment is then made in terms of: (1) standard deviation (STD) of the SSH for selected locations at each SA pass, (2) root mean square errors (RMSE) between the satellite and reference data, (3) inter-comparisons of SAMOSA2 and SAMOSA+ derived SSH results.

2 Theoretical Background

2.1 Satellite Altimetry

The SA is designed to measure the sea level by a nadir pointing radar and positioning techniques. The measurement procedure involves a pulse of radiation with known power that is transmitted from the satellite towards the sea. The pulse interacts with the rough sea surface and part of the incident radiation within the altimetric footprint reflects back to the radar altimeter, which records the returned echo of the pulse. The initial solution of the radar is limited and there is a need to process and analyse this echo by combining the observations with actual sea surface elevations over the overflight zone (i.e. Brown model for the oceanic surfaces) to get an optimal estimate of the distance. This process is called (physical fitting) retracking, it specifies the following geophysical parameters: the distance between the satellite and the subsatellite surface point (range), the significant wave height, the backscatter coefficient and the wind speed (Stammer and Cazenave 2017). The range $R(\lambda,\varphi,t)$ at a location (λ,φ) and time instant t must be corrected for the effects of the atmosphere as it delays the signal as well as for instrumental and orbital errors. The corrected range R_{corr} *is* thus represented as:

$$\mathrm{R}_{corr}\,(\lambda,\varphi,t)=\mathrm{R}\,(\lambda,\varphi,t)+corrs \tag{1}$$

where constituents (their names are rather self-explanatory, see Eq. 2) of the *corrs* term are as follows (see e.g. EUMETSAT 2018):

$$\begin{aligned} corrs = {} & \textit{sea state bias} + \textit{pole tides} + \textit{solid earth tides} \\ & + \textit{ionospheric correction} + \textit{dry tropospheric correction} \\ & + \textit{wet tropospheric correction} \end{aligned} \tag{2}$$

Note that effects of inverse barometry and ocean tides are usually included in the HDM (see Sect. 2.3). The SA derived sea surface height SSH_{SA} can then be determined by:

$$\mathrm{SSH}_{\mathrm{SA}}\,(\lambda,\varphi,t)=\mathrm{h}_{\mathrm{sat}}\,(\lambda,\varphi,t)-\mathrm{R}_{corr}\,(\lambda,\varphi,t) \tag{3}$$

where $h_{sat}(\lambda,\varphi,t)$ is a satellite altitude with respect to the geodetic reference ellipsoid.

Satellite altimetry enables studies on variable component of the SSH. For instance, the instantaneous Dynamic Topography (DT) that is SSH deviation from the geoid at the time instant t can be computed as:

$$\mathrm{DT}\,(\lambda,\varphi,t)=\mathrm{SSH}_{\mathrm{SA}}\,(\lambda,\varphi,t)-\mathrm{N}\,(\lambda,\varphi) \tag{4}$$

where is $N(\lambda,\varphi)$ is a (static) geoid height, a separation of the mean seal level related equipotential surface from the ellipsoid. Knowledge of DT behaviour (e.g. from a HDM or TG) can assist in satellite altimetry and geoid model validations.

The new ESA developed Sentinel-3A SAR altimeter offers many potential enhancements over conventional altimetry, especially in coastal areas. According to Cipollini et al. (2017) and Bonnefond et al. (2018), the new Doppler altimetry with its higher along-track resolution improves the accuracy, which becomes apparent especially in coastal zones. The way the range is obtained by SRAL altimeter differs also from the conventional altimetry functions. SAR altimetry considers the Doppler effect next to the measured time introducing change of frequency of the signal within a footprint as the satellite moves (Stammer and Cazenave 2017). The validation of the Sentinel-3A data is performed by comparison of SRAL derived SSH_{SA} (originating from two retrackers, see Sect. 3.2) with the TG corrected HDM, both referring to the geoid.

2.2 Tide Gauges

The TG measurements are referred to the zero level of the national vertical datum. Usually a set of high-precise GNSS-levelling points is used to fit the geoid model to the vertical datum. Hence, the TG derived sea surface height SSH_{TG} can be retrieved in the following way:

$$SSH_{TG}(\lambda_{TG}, \varphi_{TG}, t) = N(\lambda_{TG}, \varphi_{TG}) + DT_{TG}(t) \quad (5)$$

where $DT_{TG}(t)$ is the TG measured sea level with respect to the zero level of the national vertical datum. The TG series also need to be corrected for the effect of vertical land motion (VLM), e.g. postglacial land uplift (see Sect. 3.3). TG records are usually obtained in the near-shore. This implies that TG-s data could be under unfavourable conditions not suitable for offshore. Instead, a reliable HDM can be utilised offshore.

2.3 Hydrodynamic Models

Regionally computed HDM provide sea surface elevation DT_{HDM} as a spatio-temporal function with respect to some height reference surface, in an ideal case it would be a regional geoid model. Thus, SSH can be computed anywhere offshore (e.g. at a virtual station – VS, a location along a SA track) as follows:

$$SSH_{HDM}(\lambda_{VS}, \varphi_{VS}, t) = N(\lambda_{VS}, \varphi_{VS}) + DT_{HDM}(\lambda_{VS}, \varphi_{VS}, t) \quad (6)$$

The HDM is of particular importance when evaluating performance of Sentinel-3A SRAL as it is the only independent source for the offshore SSH. However, due to existing bias in DT_{HDM} (see Lagemaa et al. 2011), there is a need to correct the SSH_{HDM} with TG records, which yields to adjusted HDM data. Note that such regional HDM-s are constructed with respect to some seasonal mean sea level (Lagemaa 2012) which depends on the predicted amount of the seasonal water masses in sea and can therefore deviate from the historic mean sea level (and thus also from the vertical datum). Besides, the HDM models may contain a low-frequency error, the so called "zero drift" (Lagemaa et al. 2011). This spatio-temporal bias between TG and HDM data can be established at nearby TG location at the time t:

$$bias_{TG/HDM}(t) = SSH_{TG}(\lambda_{TG}, \varphi_{TG}, t) - SSH_{HDM}(\lambda_{TG}, \varphi_{TG}, t) \quad (7)$$

This bias is corrected for each corresponding cell node of the HDM grid when computing SSH_{HDM}:

$$SSH_{HDMcorr}(\lambda_{VS}, \varphi_{VS}, t) = SSH_{HDM}(\lambda_{VS}, \varphi_{VS}, t) + bias_{TG/HDM}(t) \quad (8)$$

The SSH differences at the VS between SA and reference data (HDM + TG) can then be obtained by subtracting Eq. (8) from Eq. (3):

$$SSH_{diff}(\lambda_{VS}\varphi_{VS}t) = SSH_{SA}(\lambda_{VS}\varphi_{VS}, t) - SSH_{HDMcorr}(\lambda_{VS}\varphi_{VS}t) \quad (9)$$

These expressions are on the basis for the evaluation of the Sentinel SRAL retrackers performance over the selected study area.

3 Case Study

3.1 Study Area

The performance of Sentinel-3A SRAL altimeter retrackers was evaluated in the coastal area of the Gulf of Finland (located in the eastern section of the Baltic Sea). The Gulf is a medium-size (covering approximately 30,000 km^2) sub-basin connected with the Baltic Proper and stretching to the eastern end of the Baltic Sea. The Gulf has an elongated shape, with a length of 400 km and width of 48–135 km. Due to its semi-enclosed form, the oceanographic processes are not as complex as it might be in other coastal areas, however, persistent marine features are present. The Sentinel-3A derived SSH_{SA} in this area over the duration of 2017 were compared with the ground truth that is represented by TG records in conjunction with HDM data. Three ascending (i.e. going from coast toward offshore) Sentinel-3A passes (several cycles for each) nearby three TG sites: #186 near Pirita, #300 near Loksa, #414 near Kunda in Estonia (see Fig. 1) were studied. In order to obtain comparable results, only such passes that are within 50 km from the available tide gauge data are included. Note also that retrackers show different performance for ascending or descending passes. Thus, different behaviours of altimeters errors are present which may entail various adjustments (see e.g. Mertikas et al. 2011). To avoid the discrepancy between the ascending and descending passes, this study focuses on the ascending passes only, hence applying a uniform methodology.

3.2 Satellite Altimetry

In this study the Sentinel-3A NTC (Non-Time Critical) L2 20 Hz data generated with the SAMOSA retrackers were examined. Both the SAMOSA2 and SAMOSA+ data products were retrieved from GPOD (Grid Processing on Demand) service (https://gpod.eo.esa.int), see also description of L1b products processing in Dinardo (2013). These enhanced data are to be used both in open oceans and coastal areas, however, SAMOSA+ is better tailored for coastal areas (see for example Bonnefond et al. 2018). In general, different approaches for waveform modelling are used for the

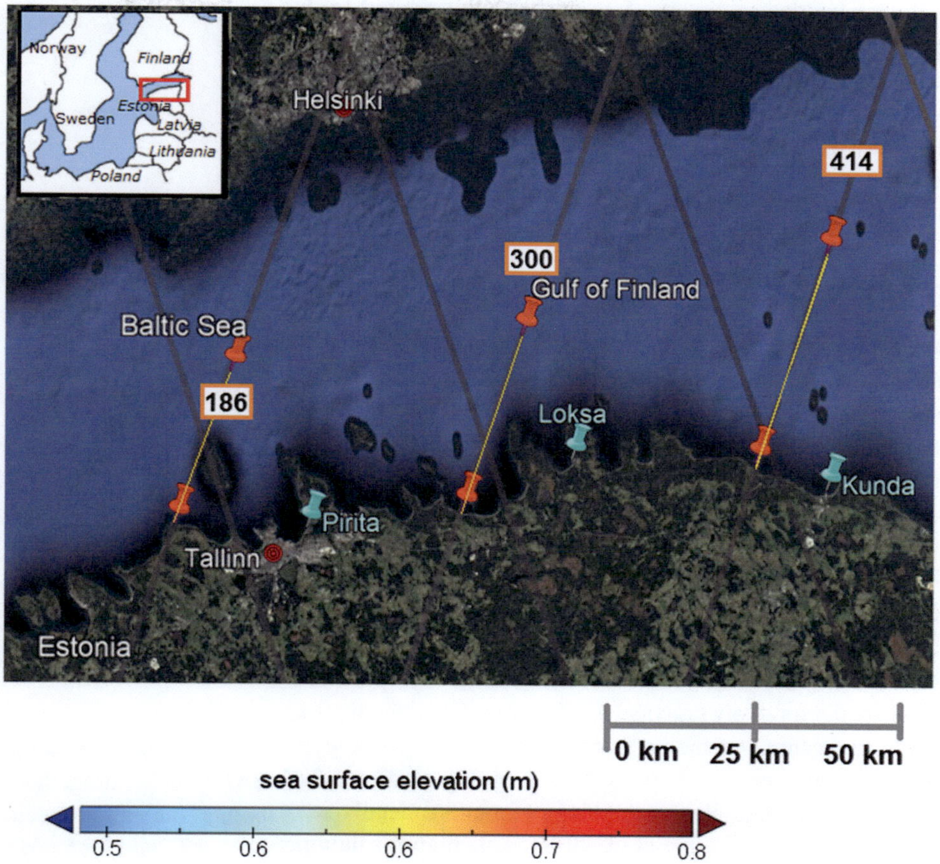

Fig. 1 Location of the study area with the sea surface elevations extracted from HDM for 21.06.2017 at 13.00 GMT, orange lines with the endpoint markers denote locations of the SSH comparison profiles, blue markers denote tide gauge sites used in this study, the pass numbered grey lines are Sentinel-3A passes

SAMOSA+ and SAMOSA2 retrackers (see a comparative description in SAMOSA Team 2013; Dinardo et al. 2015; Ray et al. 2015). Note that the SAMOSA2 data products can also be obtained from Copernicus data hub. These data were used in the initial stage of our study. It was observed, that the Copernicus retrieved SAMOSA2 data may numerically differ from the corresponding GPOD data. Apparently, the retracker is not the only difference between the two sources, distinct correction models might be applied and different waveform pre-processing is applied. Since the Copernicus data hub does not provide the SAMOSA+ data products then to compare both retrackers appropriately we use the same source – GPOD. The SSH_{SA} were calculated by Eq. (3).

3.3 Tide Gauges and Hydrodynamic Model

The TG data (provided by the Estonian Environment Agency, S. Süld, *pers. comm.*, 11/2017) are averaged to every full hour and refer to Baltic 1977 height system (i.e. the Kronstadt TG zero reading). Typical setup, data acquisition and processing principles of such pressure sensors based TG-s are described in Liibusk et al. (2013). The provided TG records are referred to the national vertical datum fitted geoid model. The entire Fennoscandia is affected by apparent land uplift at the velocity rate up to +9 mm/year, primarily due to the viscoelastic response of the solid Earth resulting from the de-glaciation of the Pleistocene ice-sheets. Over a time span this causes notable distortions of TG time series. The land uplift ratios at the three used tide gauges vary from 1.5 to 2.5 mm. Therefore, the land uplift corrections are also taken into account in sea level series by using a regional land uplift model NKG2006LU (Ågren and Svensson 2007). Three-dimensional High Resolution Oceanographic Model of the Baltic Sea, HIROMB (Funkquist2007) was used for extracting the SSH_{HDM} data. The HIROMB-BOOS Model (to be referred more generically – HDM, see Sect. 2.3) is currently used for operational sea forecast in Estonia and plays crucial role in short-term prediction of the sea level in marginal seas and coastal areas. The forecast enables a proper handling of oil spills, storm surges, navigation issues. It accounts for the contribution of the tides, currents, temperature, salinity, inverse barometric effect and sea ice coverage (Lagemaa et al. 2011). The HDM was retrieved from the Department of Marine Systems at TUT via Estonian Marine Information System, http://emis.msi.ttu.ee. The temporal resolution of the DT_{HDM} values is 1 h (resulting in 24 layers in the daily HDM), the spatial resolution is $1' \times 5/3'$ (corresponding to 1.8×1.5 km). The SSH_{HDM} values were computed by Eq. (6). For every VS we compared two sources of the SSH data (embedded in Eq. 9): SA data and the ground truth represented by the TG corrected HDM. The model bias (see Sect. 2.3) was determined by using TG stations and eliminated from the HDM model data by Eqs. (7 and 8). The standard deviation (STD) of the TG series and HDM hourly estimates vary within 4 ... 6 cm daily.

3.4 Regional Geoid Model

In 2011, after a careful and versatile revision of results of existing gravity surveys and using a satellite – only GGM (Pail et al. 2011) an enhanced Estonian gravimetric geoid model GRAV-GEOID2011 (Ellmann et al. 2011) was computed by the least squares modification of Stokes's formula (see also Ellmann 2005). That model was fitted (mean bias removed, whereas the remaining residual tilt was removed by using a polynomial fitting) to a set of 114 high-precision GPS-levelling points that resulted with the EST-GEOID-2011 model. The post-fitting residuals yielded an accuracy of 1.3 cm, which indicates the suitability of the EST-GEOID-2011 model for many practical applications. Over the marine areas its accuracy remains largely unknown, even though dedicated air- and sea-borne GNSS profiling studies exist over the West–Estonian archipelago (for more details see Gruno et al. 2013; Varbla et al. 2017). This high-resolution 1′ × 2′ (1.8 × 1.8 km) geoid model EST-GEOID2011 (Ellmann et al. 2011) was widely used by the surveying industry of Estonia as "the official national geoid" for converting the GNSS-derived heights into normal (sea level related) heights. Accordingly, the TG measurements and the TG-corrected HDM data refer to the EST-GEOID2011 model.

4 Evaluation Results

Virtual stations with values from regional geoid model and HDM were interpolated to the locations of SA measurements. Both the SAMOSA2 and SAMOSA+ retrackers provide 20 Hz data (whereas 1 Hz corrections, see Eq. (2), are available at 7 km interval on each SA track see Fig. 2). The SA measurements exceeding the predefined threshold (a difference with the ground truth data of more than 40 cm, defined as an approximate threefold value of STD) were considered being erratic, these obvious outliers were removed from further analysis.

The RMSE, STD and mean values of the SSH differences between SA SSH and the ground truth SSH (national geoid + sea elevation from TG-corrected HDM) are calculated using classical formulae. Table 1 contains the SSH_{diff} profile output statistics for SAMOSA2 and SAMOSA+ calculated for all profiles. The overall RMSE of differences (based on all SA

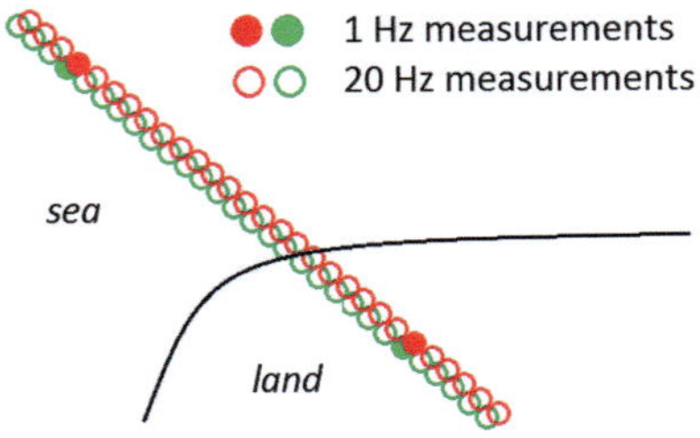

Fig. 2 The layout of SA points within a profile. Red and green colour denote the consecutive cycles

Table 1 The initial statistics of the Sentinel – 3A SAMOSA2 and SAMOSA+ retrieved overall SSH differences with the ground truth for all profiles (pass 186 cycles 17, 18, 19, 25, pass 300 cycles 18, 21 and 25, pass 414 cycles 15, 20)

Statistical characteristics	SAMOSA+ results [cm] (altogether 596 points)	SAMOSA2 results [cm] (altogether 807 points)
Overall RMSE	9.9	11.5
Overall STD	9.6	11.2
Overall mean	2.3	2.6

points within those eight profiles) is 115 and 99 mm for the SAMOSA2 (altogether 807 points) and SAMOSA+ (596 points) outputs, respectively. The overall mean of SSH_{diff} is 26 and 23 mm for SAMOSA2 and SAMOSA+, respectively. The obtained RMSE (99 and 115 mm) values suggest that SAMOSA+ output is more reliable than SAMOSA2 data for the given study area and chosen time period. This is not surprising, since SAMOSA+ is claimed to perform better in the coastal area (see e.g. the Bonnefond et al. 2018 study over Mediterranean Sea) as it is tailored specifically for coastal zones and other particular areas (inland waters, sea ice). SSH differences between SAMOSA2 and SAMOSA+ for three chosen SA passes are compared in Fig. 3. The statistics for detected discrepancies of SSH obtained from SAMOSA+ referring to all computed individual passes and cycles are presented in Table 2. The SAMOSA+ retracker obtained SSH differences are shown in Fig. 4 for three selected passes. Similarly, the SAMOSA2 output related values are represented in Table 3 and Fig. 5 for three chosen passes as well.

In general SAMOSA+ and SAMOSA2 retrackers are able to retrieve data of similar quality close to the coast (see Fig. 3). It can be also assumed that RMS values are somewhat affected by the fact that less data was available close to the coast or was labelled as outliers. This affects RMS, a similar observation was made by Bonnefond et al. (2018).

Notice that profile lengths are different, even for the same pass but different cycles. This is due to different number of outliers removed or different amount of missing or not reliable enough information provided (e.g. atmospheric

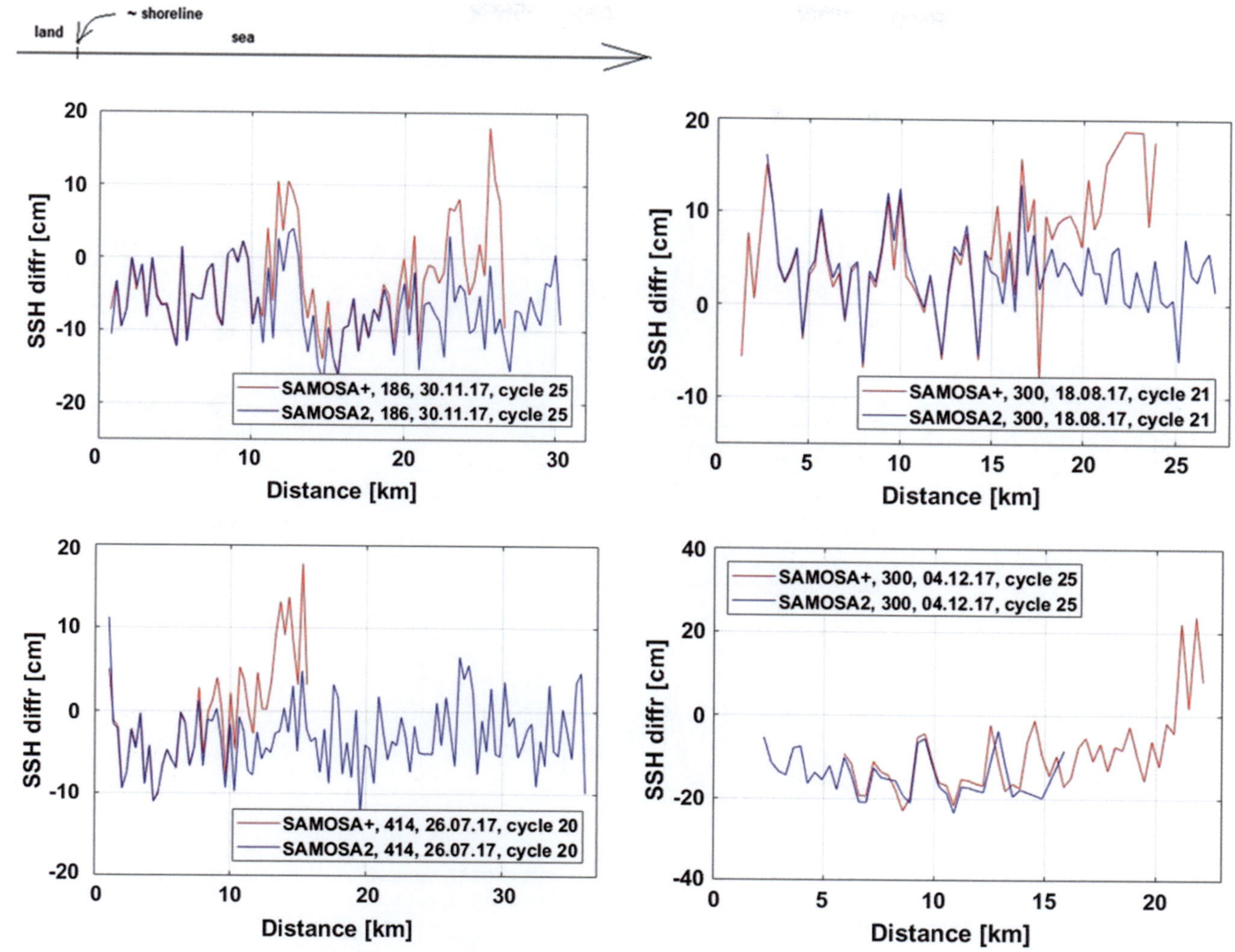

Fig. 3 Examples of comparisons of detected discrepancies between the SAMOSA2 and SAMOSA+ derived SSH-s with the "ground truth" (geoid + TG-corrected HDM, represented by the zero values) for four selected overflights along three different Sentinel-3A tracks (for statistics refer to Tables 2 and 3). The horizontal axis denotes the distance from the beginning of the profile (the closest point to the shoreline, located in the left-hand side of the figure). Note the different vertical scale for each sub-plot

Table 2 Statistics of the Sentinel-3A SAMOSA+ retrieved SSH differences with the ground truth

Pass	Cycle	No of 20 Hz points	Profile length [km]	Distance of shoreline to the closest profile-point [km]	RMSE [cm]	STD [cm]	Mean [cm]
186	17	88	29.8	0.5	9.7	7.3	−6.3
	19	91	29.8	0.3	10.1	5.5	8.5
	25	79	25.8	0.5	7.5	6.7	−3.5
300	18	81	27.1	1.4	11.2	8.4	7.5
	21	65	22.5	1.5	8.1	6.1	5.4
	25	50	16.2	7.7	13.5	9.3	−9.9
414	15	107	35.1	0.8	10.7	5.8	9.0
	20	45	14.5	1.0	6.4	6.5	−0.2

corrections). This limitation has a significant impact on the resultant statistics and thus the presence of initially existing (and later excluded) outliers will be examined in future studies. We consider this appropriate to not bring the profiles' lengths together as that would mean ignoring good enough quality data (after outlier removal) close to the coast

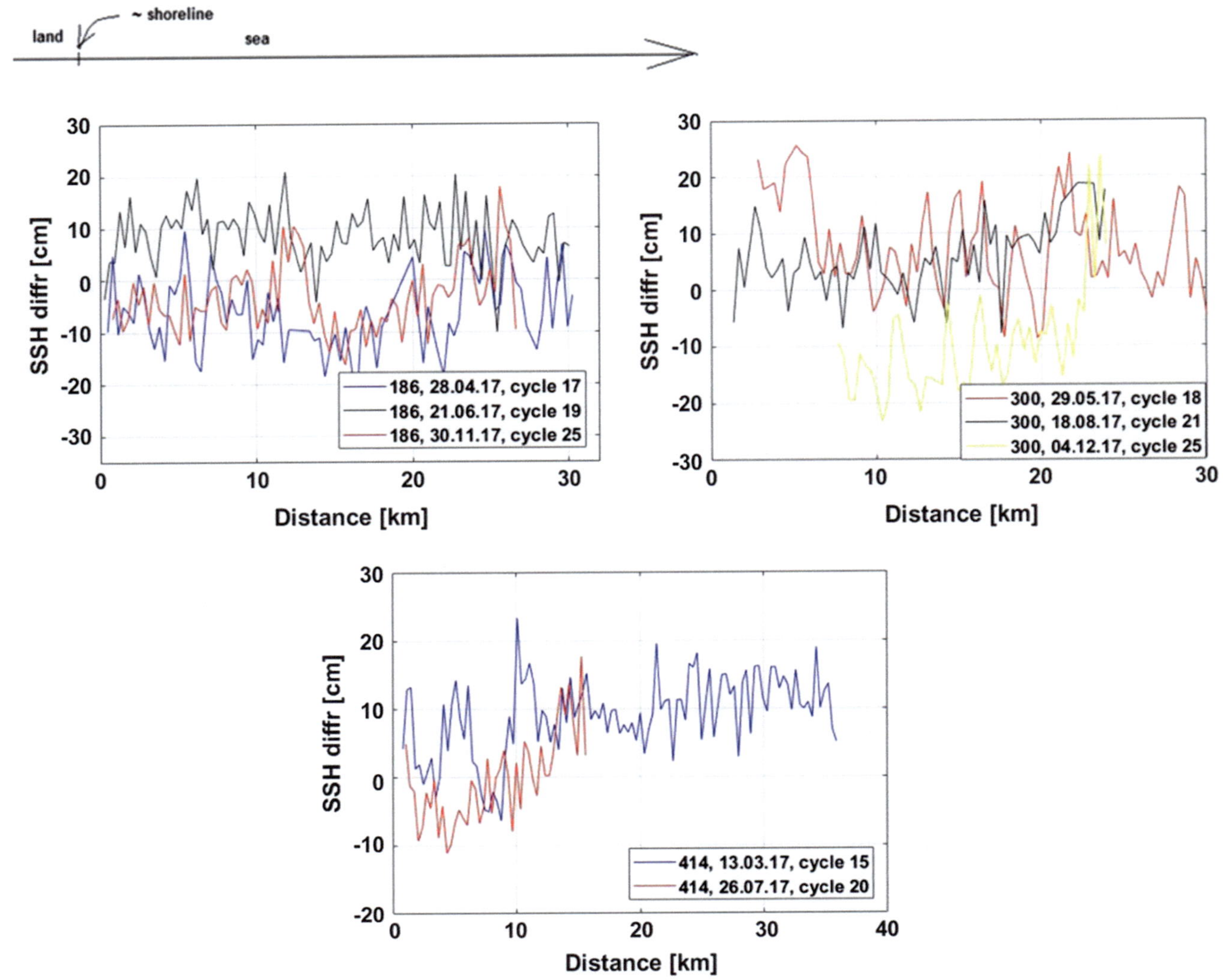

Fig. 4 The detected discrepancies between the SAMOSA+ derived SSH and "ground truth" (geoid and TG-corrected HDM) along the three selected Sentinel-3A passes. The horizontal axis denotes the distance from the beginning of the profile (the left-hand side)

Table 3 Statistics of the Sentinel – 3 SAMOSA2 retrieved SSH differences with the ground truth

Pass	Cycle	No of 20 Hz points	Profile length [km]	Distance of shoreline to the closest profile-point [km]	RMSE [cm]	STD [cm]	Mean [cm]
186	17	72	24.5	5.7	10.3	7.7	−6.9
	19	91	29.8	0.2	19.9	5.4	19.1
	25	90	29.4	0.6	8.2	4.9	−6.6
300	18	81	27.1	1.4	12.6	9.6	8.2
	21	75	24.5	3.9	5.4	4.1	3.5
	25	36	16.6	2.3	15.4	5.1	−14.5
414	15	107	35.1	1.0	10.7	5.8	9.0
	20	107	35.1	1.0	5.2	4.2	−3.2

where retrackers' performance appear different. For profiles validation it was necessary to include the HDM as it provides the information about the offshore SSH. That implies that the results are affected by errors stemming from that model in addition to altimeter errors. This is very important for validation of SA data as knowledge of uncertainty of the model enables better interpretation of the differences between SA and ground truth data.

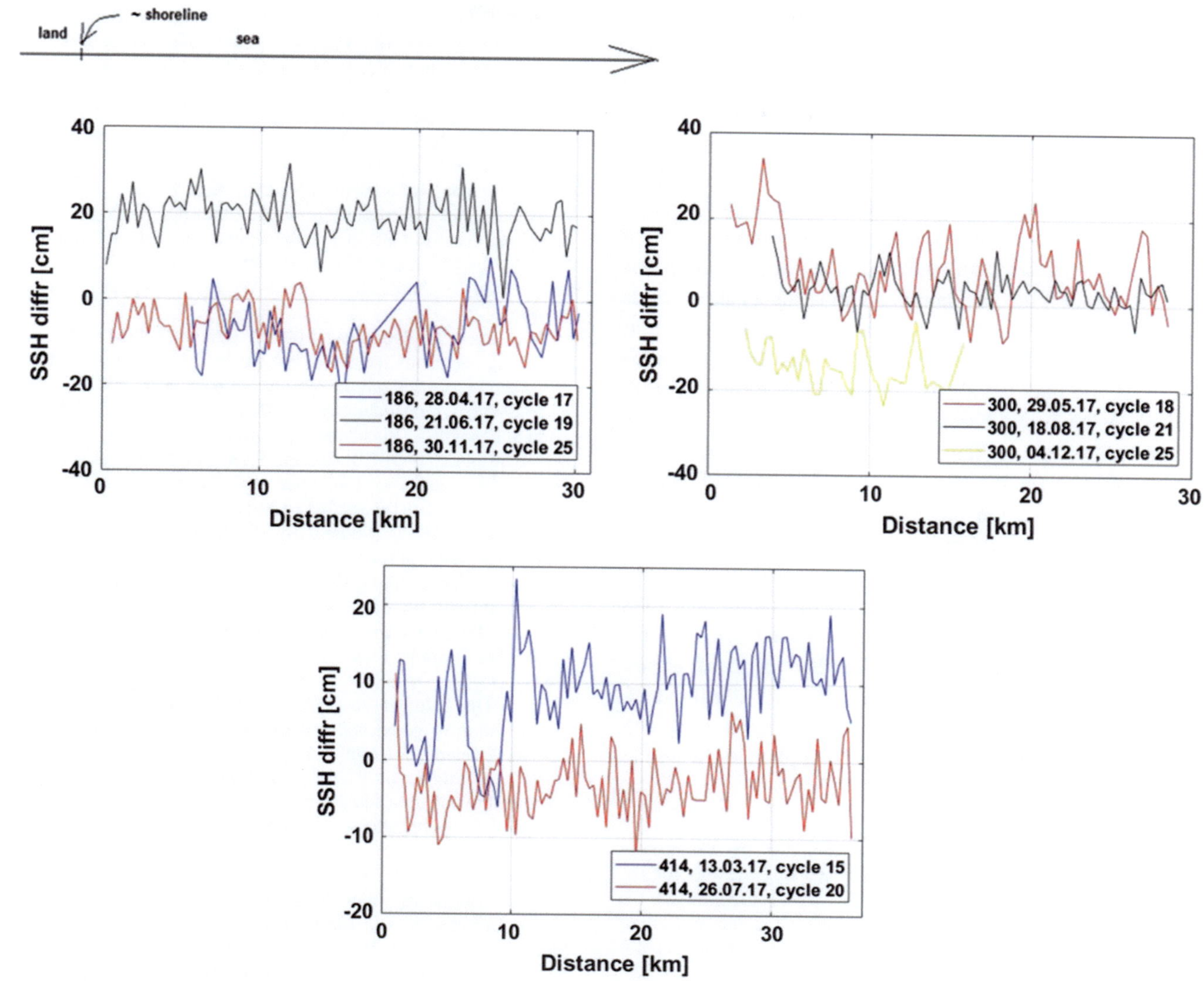

Fig. 5 The detected discrepancies between the SAMOSA2 derived SSH and "ground truth" (geoid and TG-corrected HDM) along the three selected Sentinel-3A passes. The horizontal axis denotes the distance from the beginning of the profile (the left-hand side)

It should be noted that for this study the emphasis is more on the methodology employed than the interpretation of the contribution of this error. However, this error will be taken into account when analysing the results. The results are influenced also by other factors, among others by track direction, season, sea state.

The above results confirm the assumptions that Doppler altimetry enhances the quality of the SSH_{SA} compared to conventional altimetry (the performance assessments can be found e.g. in Cheng 2004; Dong et al. 2002; Kolenkiewicz and Nerem 1994), but complexity of the Baltic Sea contributes adversely to it. RMS errors of discrepancies obtained in this study are better than as evaluated in other regions in case of SAMOSA2 retracker output (Passaro et al. 2015; Bonnefond et al. 2003).

5 Conclusions and Further Studies

This study examined the performance of the Sentinel-3A SRAL in the coastal areas of the Baltic Sea using two retrackers (SAMOSA2 and SAMOSA+). The use of the local geoid model contributed significantly to validation of SA data due to its good quality, serving as a correction to the SA and reference data to refer them to the same vertical datum. The obtained differences in SSHs and overall RMS error of 115 mm for SAMOSA2 and 99 mm for SAMOSA+ yield better agreement of SA with the ground truth than conventional altimetry in Baltic Sea (Madsen et al. 2007). Based on the methodology applied, the results suggest that in all cases the SAMOSA2 output is less reliable than its upgrade SAMOSA+. For explanations on such discrepancies it requires further examination into: (1) methodology applied for the waveform processing in the retrackers, (2) the method used in applying HDM virtual stations that coincide with satellite points and (3) local conditions of the study area. The results strongly suggests whilst available retrackers are capable of increasing the accuracy of satellite derived SSH, further research is required in development of a customized retracker taking into account sensitive conditions of the Baltic Sea such as land contamination by the various archipelagos that exists, local atmospheric conditions that can be unpredictable and coastal processes that may influence SSH and may not always be predicted accurately with models (Delpeche-Ellmann et al. 2017, 2018).

The results of the Sentinel-3A SRAL observations are overall very promising. For instance, in terms of marine engineering application the results can be integrated into the international collaboration FAMOS project (Finalizing surveys for the Baltic motorways of the sea), which aims at enhancing marine geoid model over the Baltic Sea. This study shows great potential in using coastal SA data for bridging the gap in between the on-land geodetic infrastructure and the open sea marine products (e.g. Märdla et al. 2017; Ellmann et al. 2019). The developed methodology with some improvements can presumably be applied elsewhere in near-coast marine areas.

Acknowledgements This research was supported by Connecting Europe Facility (CEF) project "FAMOS (Finalising Surveys for the Baltic Motorways of the Sea) Odin" (project VEU16013) and TUT's internal funding project SS440 "Iterative marine geoid modelling in near-coast regions by using re-tracked satellite altimetry, in-situ and modelled data". The participation of the prime author in the "International Review Workshop on Satellite Altimetry cal/val activities and applications" in 26-26.04.2018 was supported by International Association of Geodesy. We thank two anonymous reviewers whose helpful comments improved this manuscript.

References

Ågren J, Svensson R (2007) Postglacial land uplift model and system definition for the New Swedish Height System RH 2000. LMV-rapport 2007:4, Lantmäteriet, Sweden

Bonnefond P, Exertier P, Laurain O, Menard Y, Orsoni A, Jan G, Jeansou E (2003) Absolute calibration of Jason 1 and TOPEX/Poseidon altimeters in Corsica, special issue on Jason 1 calibration/validation, part 1. Mar Geod 26(3–4):261–284. https://doi.org/10.1080/714044521

Bonnefond P, Laurain O, Exertier P, Boy F, Guinle T, Picot N, Labroue S, Raynal M, Donlon C, Féménias P, Parrinello T, Dinardo S (2018) Calibrating the SAR SSH of Sentinel-3A and CryoSat-2 over the Corsica facilities. Remote Sens 10:92. https://doi.org/10.3390/rs10010092

Cheng K-C (2004) Radar altimeter absolute calibration using GPS water level measurements. Geodetic and GeoInformation Science Department of Civil and Environmental Engineering and Geodetic Science, The Ohio State University Columbus, Ohio

Cipollini P, Calafat FM, Jevrejeva S et al (2017) Monitoring Sea level in the coastal zone with satellite altimetry and tide gauges. Surv Geophys 38:33. https://doi.org/10.1007/s10712-016-9392-0

Delpeche-Ellmann N, Mingelaitė T, Soomere T (2017) Examining Lagrangian surface transport during a coastal upwelling in the Gulf of Finland, Baltic Sea. J Mar Syst 171:21–30. https://doi.org/10.1016/j.jmarsys.2016.10.007

Delpeche-Ellmann N, Soomere T, Kudryavtseva N (2018) The role of nearshore slope on cross-shore surface transport during a coastal upwelling event in Gulf of Finland, Baltic Sea. Estuar Coast Shelf Sci 209:123–135. https://doi.org/10.1016/j.ecss.2018.03.018

Dinardo S (2013) Guidelines for the SAR (Delay-Doppler) L1b processing

Dinardo S, Lucas B, Benveniste J (2015) Sentinel-3 STM SAR ocean retracking algorithm and SAMOSA model. In: 2015 IEEE international geoscience and remote sensing symposium (IGARSS), Milan, pp 5320–5323. https://doi.org/10.1109/IGARSS.2015.7327036

Dinardo S, Fenoglio L, Buchhaupt C, Becker M, Scharroo R, Fernandes MJ, Benveniste J (2017) Coastal SAR and PLRM altimetry in German Bight and West Baltic Sea. Adv Space Res 62:1371–1404. https://doi.org/10.1016/j.asr.2017.12.018

Dong X, Woodworth PL, Moore P, Bingley R (2002) Absolute calibration of the TOPEX/Poseidon altimeters using UK tide gauges, GPS and precise, local geoid-differences. Mar Geod 25:189–204. https://doi.org/10.1080/01490410290051527

Ellmann A (2005) Two deterministic and three stochastic modifications of Stokes's formula: a case study for the Baltic countries. J Geod 79:11–23

Ellmann A, Oja T, Jürgenson H (2011) Application of space technologies to improve geoid and gravity field models over Estonia (in Estonian). Geodeet 41(65):22–25

Ellmann A, Märdla S, Oja T (2019) The 5 mm geoid model for Estonia computed by the least squares modified Stokes's formula. Surv Rev. Accepted for publications. https://doi.org/10.1080/00396265.2019.1583848

EUMETSAT (2018) Sentinel-3 SRAL marine user handbook

Gruno A, Liibusk A, Ellmann A, Oja T, Vain A, Jürgenson H (2013) Determining sea surface heights using small footprint airborne laser scanning. In: Proceedings of the SPIE 8888, Remote Sensing of the Ocean, Sea Ice, Coastal Waters, and Large Water Regions 2013, 88880R (16 October 2013). https://doi.org/10.1117/12.2029189

Kolenkiewicz R, Nerem RS (1994) Calibration of TOPEX/POSEIDON at platform harvest. J Geophys Res 99(C12):24465–24485

Lagemaa P (2012) Operational forecasting in Estonian marine waters. PhD thesis, TUT Press

Lagemaa P, Elken J, Kõuts T (2011) Operational Sea level forecasting in Estonia. Est J Eng 17(4):301–331

Liibusk A, Ellmann A, Kõuts T, Jürgenson H (2013) Precise hydrodynamic leveling by using pressure gauges. Mar Geod 36(2):138–163. https://doi.org/10.1080/01490419.2013.771594

Madsen KS, Høyer JL, Tscherning CC (2007) Near-coastal satellite altimetry: sea surface height variability in the North Sea – Baltic Sea area. Geophys Res Lett 34(14):L14601. https://doi.org/10.1029/2007GL029965

Märdla S, Ågren J, Strykowski G, Oja T, Ellmann A, Forsberg R, Bilker-Koivula M, Omang O, Paršeliūnas E, Liepinš I, Kaminskis J (2017) From discrete gravity survey data to a high-resolution gravity field representation in the Nordic-Baltic region. Mar Geod 40(6):416–453. https://doi.org/10.1080/01490419.2017.1326428

Mertikas SP, Daskalakis A, Tziavos IN, Georgios S, Vergos GS, Xenophon Frantzis X, Achilleas Tripolitsiotis A, Partsinevelos P, Andrikopoulos D, Zervakis V (2011) Ascending and descending passes for the determination of the altimeter bias of Jason satellites using the Gavdos facility. Mar Geod 34(3):261–276. https://doi.org/10.1080/01490419.2011.584837

Pail R, Bruinsma S, Migliaccio F, Förste C, Goiginger H, Schuh W-D, Höck E, Reguzzoni M, Brockmann JM, Abrikosov O, Veicherts M, Fecher T, Mayrhofer R, Krasbutter I, Sansò F, Tscherning CC (2011) First GOCE gravity field models derived by three different approaches. J Geod 85:819. https://doi.org/10.1007/s00190-011-0467-x

Passaro M, Cipollini P, Vignudelli S, Quartly GD, Snaith HM (2014) ALES: a multi-mission adaptive subwaveform retracker for coastal and open ocean. Remote Sens Environ 145(2014):173–189

Passaro M, Cipollini P, Benveniste J (2015) Annual sea level variability of the coastal ocean: the Baltic Sea-North Sea transition zone. J Geophys Res Oceans 120(4):3061–3078

Ray C, Martin-Puig C, Clarizia MP, IEEE Member, Ruffini G, Dinardo S, Gommenginger C, Benveniste J (2015) SAR altimeter backscattered waveform model. IEEE Trans Geosci Remote Sens 53:911–919. https://doi.org/10.1109/TGRS.2014.2330423

SAMOSA Team (2013) Detailed processing model of the Sentinel-3 SRAL SAR altimeter ocean waveform retracker. ESA, v2.3.0

Stammer D, Cazenave A (2017) Satellite altimetry over oceans and land surfaces. In: Earth observation of global changes, 1st edn. Taylor & Francis, Abingdon

Varbla S, Ellmann A, Märdla S, Gruno A (2017) Assessment of marine geoid models by ship-borne GNSS profiles. Geodesy Cartography 43(2):41–49

Woodworth PL, Pugh DT, Plater A (2015) Sea-level measurements from tide gauges. In: Handbook of sea-level research Shennan/Handbook of sea-level research. Wiley, Hoboken, pp 555–574. https://doi.org/10.1002/9781118452547.ch35

Sea Level Variability in the Strait of Gibraltar from Along-Track High Spatial Resolution Altimeter Products

Jesús Gómez-Enri, Stefano Vignudelli, Alfredo Izquierdo, Marcello Passaro, Carlos José González, Paolo Cipollini, Miguel Bruno, Óscar Álvarez, and Rafael Mañanes

Abstract

Accurate coastal altimetry data are important for coastal observing systems (monitoring) and to re-analyze previous datasets. In this work, we analyzed the cross-strait variability in the eastern side of the Strait of Gibraltar using one descending track from the European Space Agency (ESA) Envisat RA-2 descending track #0360. We developed an accurate coastal altimetry product at high spatial resolution along track (∼350 m between two consecutive 18-Hz measurements). We focused on the analysis of the spatio-temporal variability of along-track Absolute Dynamic Topography (ADT) profiles. We first estimated the Sea Level Anomalies (SLA) using the Adaptive Leading Edge Subwaveform (ALES) retracker. To do this, an along-track Mean Sea Surface based on ALES data was computed by interpolating the along-track SSH profiles onto nominal tracks. Then, along-track ADTs were obtained using a local Mean Dynamic Topography (MDT) from a two-dimensional (depth-averaged), two-layer, finite-difference, hydrodynamic model (*UCA2.5D*). The cross-strait variability observed with the ADT profiles and its dependence with the wind regime was analyzed and discussed. Our preliminary results from the improved altimetry data sets confirm what was previously found in the area using only tide gauge data. The joint processing and exploitation approach can be applied to Sentinel-3A and future altimeter missions, and might be extended to other challenging coastal zones.

Keywords

Cross-strait sea level variability · Satellite altimetry · Strait of Gibraltar · Tide gauge

J. Gómez-Enri (✉) · A. Izquierdo · C. J. González · M. Bruno · Ó. Álvarez · R. Mañanes
Applied Physics Department, University of Cadiz, Cadiz, Spain
e-mail: jesus.gomez@uca.es

S. Vignudelli
CNR Institute of Biophysics (CNR-IBF), Pisa, Italy

M. Passaro
Deutsches Geodätisches Forschungsinstitut der Technischen Universität München (DGFI-TUM), Munich, Germany

P. Cipollini
Telespazio Vega for ESA Climate Office, Harwell, UK

1 Introduction

The main objective of this paper is to show that accurate sea level data can be retrieved in a challenging location such as the Strait of Gibraltar. These data can be used to improve our knowledge of the hydrodynamic processes in the Strait. We show one example of that by analyzing the cross-strait sea level variability and its relation to the wind regime. Our analysis shows that coastal altimetry data are ready for exploitation but care must be taken to fully understand the physical content of the products.

S. P. Mertikas, R. Pail (eds.), *Fiducial Reference Measurements for Altimetry*,
International Association of Geodesy Symposia 150, https://doi.org/10.1007/1345_2019_54

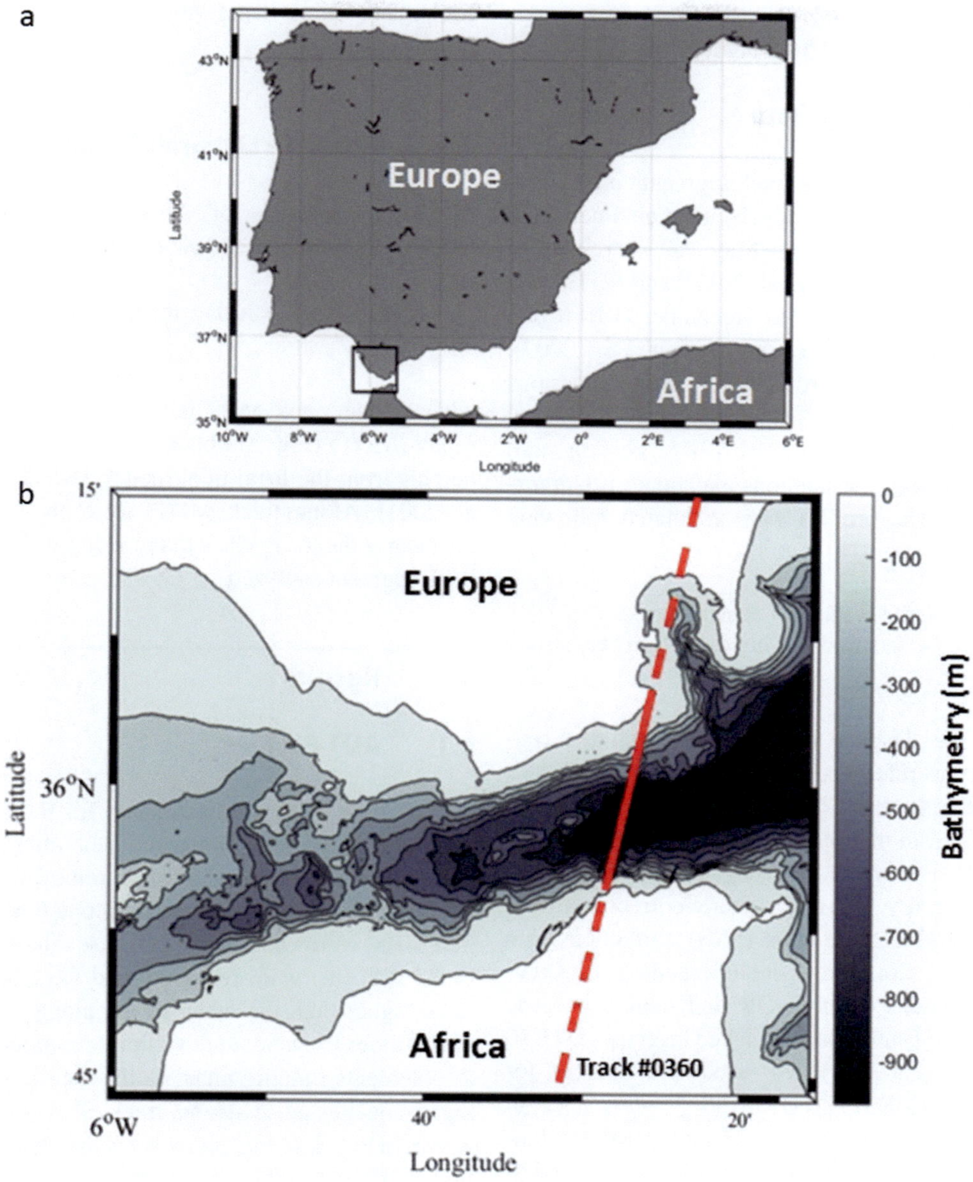

Fig. 1 Study area. The Strait of Gibraltar is the choke point between Europe and Africa (**a**). The Envisat RA-2 track segment analyzed (descending track #0360) and the bathymetry are shown in (**b**)

2 Study Area

The Strait of Gibraltar connects the Atlantic Ocean and the Mediterranean Sea (Fig. 1a). Its bathymetry is deeper in its eastern side and shallower in the western part (Fig. 1b). The different water properties, less salty and warmer Atlantic water vs. saltier and cooler Mediterranean makes a strong density-stratification in the water column. The Strait has been deeply analyzed in the last decades with in situ campaigns that collected valuable data. The hydrodynamic regime was characterized by Lacombe and Richez (1982). Many other authors contributed to a better knowledge of the physics of the Strait (e.g., Candela et al. 1989; Garrett et al. 1990; Brandt et al. 2004; Stanichny et al. 2005; Hughes et al. 2015). From the three scales of variability in the Strait we will focus on the subinertial scale, ranging from days to months and accounting for the exchange flows due to meteorological forcing (Candela et al. 1989).

3 Datasets and Methodology

3.1 Satellite Altimetry Data

High-rate (18 Hz) data along the track segment corresponding to descending track #0360 (Fig. 1b) were obtained from two sources: (1) ERS-2 RA: from May 1995 (cycle 1) to June 2003 (cycle 85) (Brockley et al. 2017); and (2) Envisat RA-2: from May 2002 (cycle 6) to September 2010 (cycle 93) corresponding to Phase E2 of the mission (ESA 2007). Sensor Geophysical Data Records (SGDR) of both missions were provided by the European Space Agency (ESA). Along track sea level anomalies (SLA) (at 18-Hz posting rate) were obtained following the methods explained in Gómez-Enri et al. (2016). The profiles were estimated following Eq. (1):

$$\text{Along\_track\_SLA} = \text{Orbit} - \text{Range} - \text{Range Correc.} - \text{Geophysical Correc.} - \text{Mean Sea Surface} \quad (1)$$

where Orbit is the distance between the satellite's center of mass and the reference surface (ellipsoid WGS84), which is available in the SGDR product; Range is the retracked Range from the Adaptive Leading Edge Subwaveform: ALES (Passaro et al. 2014); Range Corrections include ionospheric, dry, wet tropospheric corrections from SGDR; as well as Sea State Bias (SSB) correction from ALES; Geophysical Corrections include solid earth, geocentric pole corrections from SGDR and total geocentric ocean tides from the Danmarks Teknishe Unversitet DTU10 (Cheng and Andersen 2011). We used two sources for Mean Sea Surface (MSS), DTU15MSS (Andersen et al. 2016), and the along-track local MSS (local_MSS) computed combining all the overpasses from ERS-2 and Envisat missions based on ALES retrievals of Range and SSB.

3.2 In Situ Data

The cross strait sea level variability was also analyzed with two pressure tide gauges located at Ceuta and Tarifa (Fig. 1b). The instruments belong to Puertos del Estado (http://www.puertos.es). The sampling recorded water levels at a 5-minutes time interval. Sea level anomalies were obtained removing the tides and the mean sea level relative to the instruments. Hourly time series of 10-m height wind speed and direction were obtained from a weather station deployed by the Agencia Estatal de Meteorología (http://www.aemet.es/es/portada) at Tarifa (Fig. 1b). The zonal component of the wind velocity (u) was estimated in the study time period.

3.3 Absolute Dynamic Topography (ADT)

Along-track profiles of Absolute Dynamic Topography (ADT) were obtained following Eq. (2):

$$\text{Alongtrack\_ADT} = \text{Along\_track\_SLA} + \text{Along\_track\_MDT} \quad (2)$$

where along_track_MDT is a local Mean Dynamic Topography (MDT) (Fig. 2) obtained using 1 year of sea surface heights from the local tidal model, UCA2.5D (Izquierdo et al. 2001). Along_track_MDTs were obtained interpolating bi-linearly the local MDT to the along-track positions of the track segment analyzed.

4 Results

4.1 ADT Profiles

Along_track_SLA were estimated for the Envisat mission. 30 out of 88 cycles were retained after data screening. We used three criteria for data removal: (1) Only radar measurements gathered in ocean mode (chirp bandwidth of 320 MHz) were considered; (2) SLA values out of the range [−2.5 2.5] (m) were removed; and (3) Absolute values of SLA higher than the mean of the along-track SLA profile plus 3 times its standard deviation were rejected. We applied a 5-elements running mean to the data in order to remove high frequency noise. Along_track_SLAs for valid cycles are shown in Fig. 3. In Fig. 3a we show the results obtained using DTU15MSS as reference surface to obtain the anomalies. The SLA values range between −0.8 and −0.2 m with a marked sea level difference between the southern (smaller) and the northern (bigger) sectors of the Strait. This is not in agreement with previous works using in situ measurements. Ross et al. (2000) showed that a higher sea level is observed in the southern side of the Strait with respect to the northern sector, with a difference ranging between 0.2 m and 0.1 m (Bormans and Garrett 1989). The use of the global mean sea level (DTU15MSS) could explain the disagreement. We obtained the SLA profiles using the ERS2/Envisat local mean sea surface as reference. The results are shown in Fig. 3b. Here, a positive cross-strait sea level difference between the southern and northern sectors is observed; this presents a better agreement with in situ observations (Ross et al. 2000). For this reason, the local mean sea surface (local_MSS) was used to estimate the profiles of ADT.

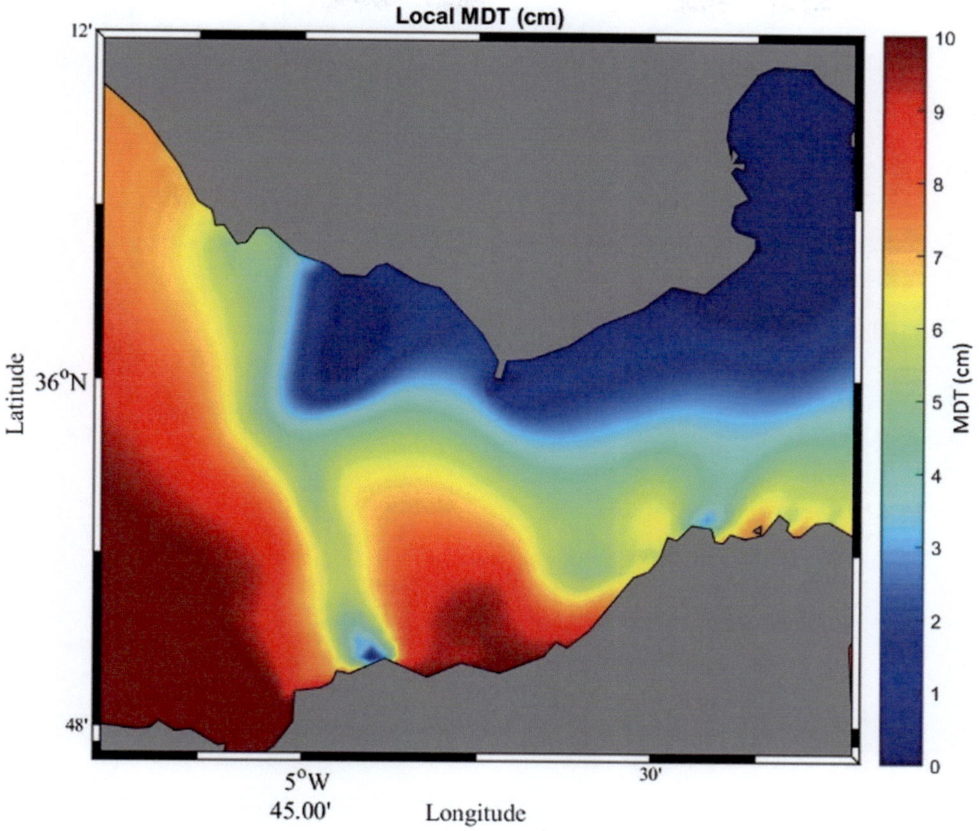

Fig. 2 Mean Dynamic Topography (in cm) used to estimate the Absolute Dynamic Topography

4.2 Sea Level Variability Due To the Wind Regime

Figure 3b reveals an inversion in the positive (south–north) cross strait sea level in some cycles. This might be related to strong easterly winds as suggested by Stanichny et al. (2005). The authors used two tide gauges at both sides of the Strait analyzing the relation of the cross-strait sea level variability with the wind regime. Figure 4 shows one example of a positive cross-strait sea level difference (Fig. 4a: cycle 93) and one example of a negative difference (Fig. 4c: cycle 47). Three days of the zonal component of the wind considering the dates of the satellite crossing the Strait is also shown in Fig. 4b (cycle 93) and Fig. 4d (cycle 47). The inversion of the cross-strait sea level difference is clearly observed in cycle 47 when a negative zonal component of the wind (easterly) was observed during more than 15 h before the satellite pass. The sea level difference between the southern and northern tide gauges was 6.6 cm (date and time corresponding to cycle 93) and − 1.0 cm (cycle 47), confirming the cross-strait sea level drops during the easterly event.

5 Final Remarks

In this study, we used accurate ADT data from the Envisat mission to analyze the cross-strait variability in the eastern side of the Strait of Gibraltar and its relation to the wind regime. Along track SLAs were obtained for the ERS2/Envisat descending orbit #0360 based on ALES reprocessing and improved corrections. The analysis was also made using two tide gauges located at the southern and northern sides of the Strait. Realistic along-track ADTs were obtained with a local along-track MSS (based on ALES). We observed positive (negative) sea level differences between the southern and northern sectors of the Strait were related to positive (negative) values of the zonal component of the wind. This was confirmed by the analysis of the outputs of two tide gauges located at Ceuta (south) and Tarifa (north). This analysis will be completed with the new generation of SAR-mode satellite altimeters (Sentinel-1A/B). Accurate sea level data in the Strait of Gibraltar will contribute to a better knowledge of the mechanisms of water exchange through it.

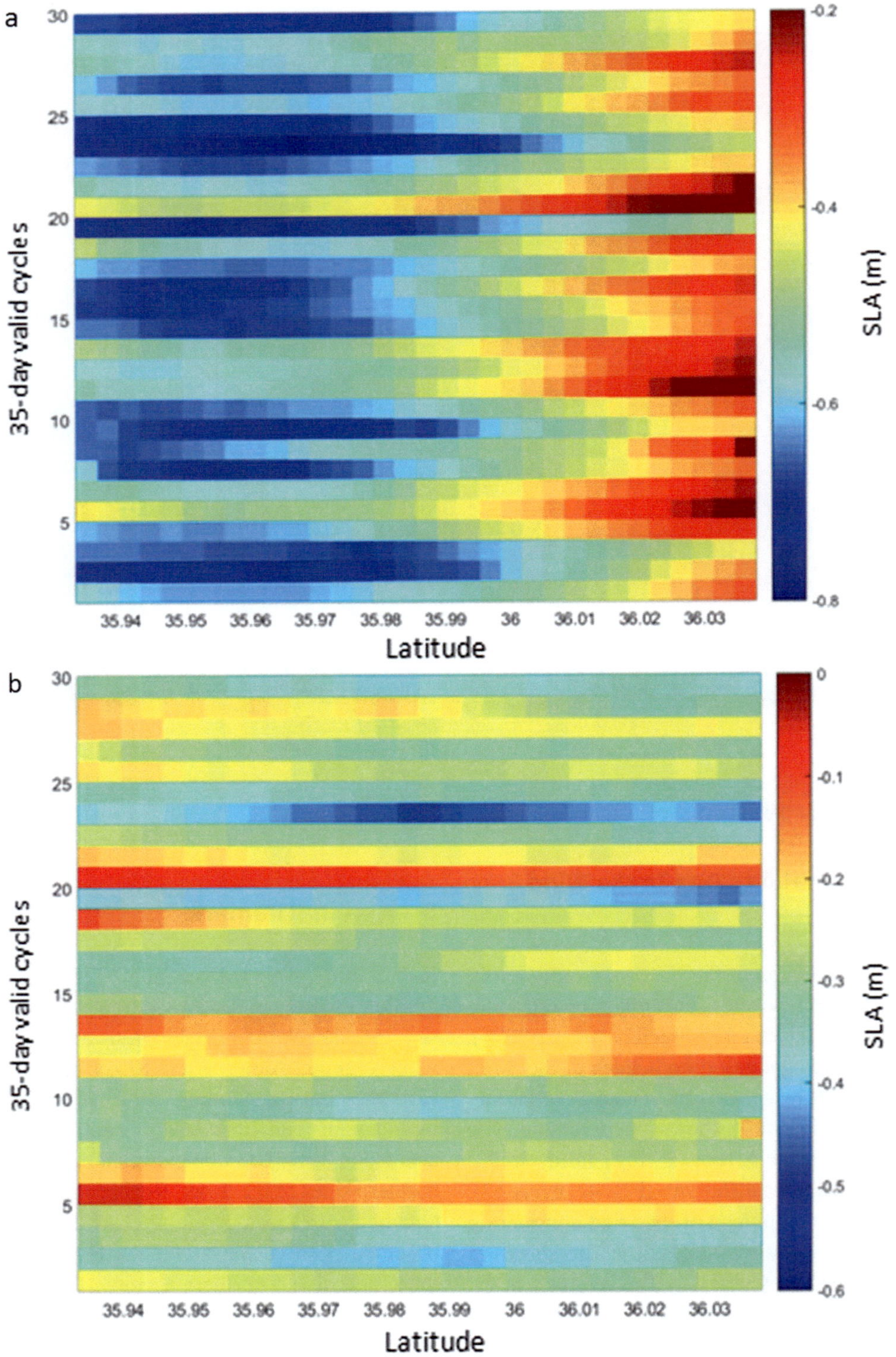

Fig. 3 Envisat RA-2 profiles of sea level anomalies using DTU15MSS (**a**) and Local_MSS (**b**) as mean sea surfaces. Only valid cycles are shown

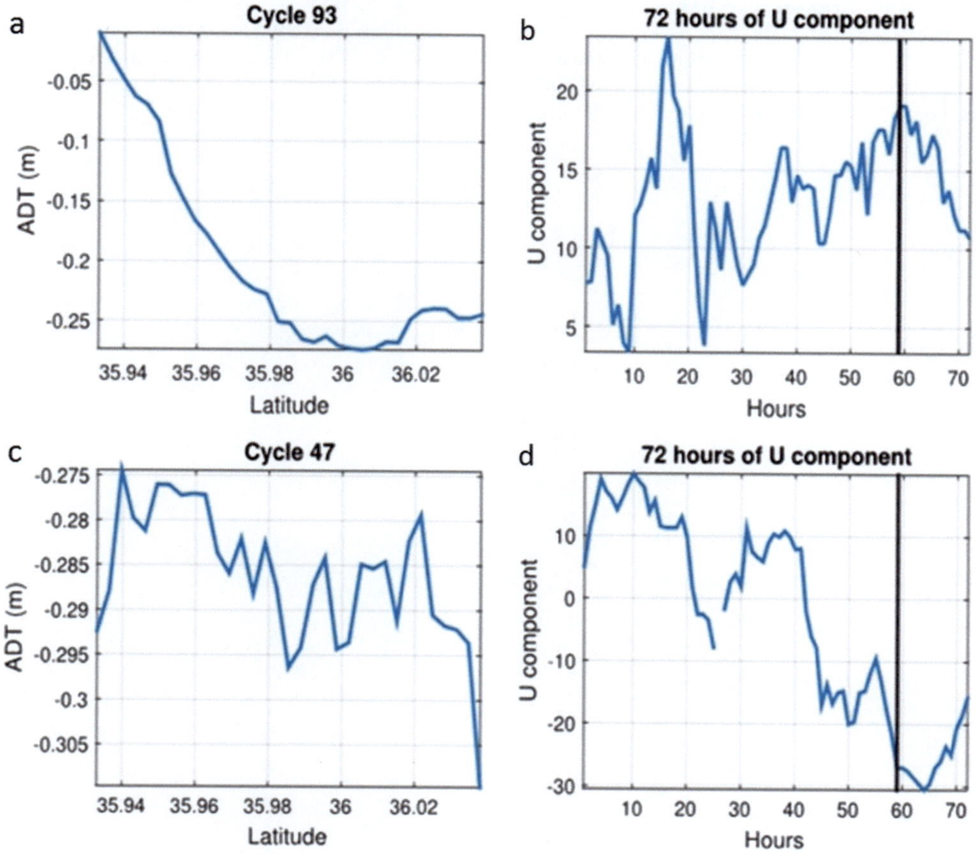

Fig. 4 ADT for cycle 93 (**a**) and for cycle 47 (**c**). 72 h of the zonal component u of the wind using as reference the date of the Envisat pass in the study area (track #0360) for cycle 93 (**b**) and for cycle 47 (**d**). The vertical black line in (**b**) and (**d**) shows the time of the Envisat passes

Acknowledgments Special thanks are given to the Spanish Instituto Hidrográfico de la Marina (bathymetry dataset), Puertos del Estado (tide gauge data), and Agencia Estatal de Meteorología (in situ wind data).

References

Andersen OB, Stenseng L, Piccioni G, Knudsen P (2016) The DTU15 MSS (Mean Sea Surface) and DTU15LAT (Lowest Astronomical Tide) reference surface. Presented at the ESA Living Planet Symposium, Prague

Bormans M, Garrett C (1989) The effects of nonrectangular cross section, friction, and barotropic fluctuations on the exchange through the Strait of Gibraltar. J Phys Oceanogr 19(10):1543–1557. https://doi.org/10.1175/1520-0485(1989)019<1543:TEONCS>2.0.CO;2

Brandt P, Rubino A, Sein DV, Baschek B, Izquierdo A, Backhaus JO (2004) Sea level variation in the Western Mediterranean studied by a numerical tidal model of the Strait of Gibraltar. J Phys Oceanogr 34:433–443. https://doi.org/10.1175/1520-0485(2004)034<0433:SLVITW>2.0.CO;2

Brockley DJ, Baker S, Féménias P, Martínez B, Massmann FH, Otten M, Paul F, Picard B, Prandi P, Roca M, Rudenko S (2017) REAPER: reprocessing 12 years of ERS-1 and ERS-2 altimeters and microwave radiometer data. IEEE Trans Geosci Remote Sens 55:5506–5514. https://doi.org/10.1109/TGRS.2017.2709343

Candela J, Winant CD, Bryden HL (1989) Meteorologically forced subinertial flows through the Strait of Gibraltar. J Geophys Res 94:12667–12674. https://doi.org/10.1029/JC094iC09p12667

Cheng Y, Andersen OB (2011) Multimission empirical ocean tide modelling for shallow waters and polar seas. J Geophys Res-Oceans 166(C11). https://doi.org/10.1029/2011JC007172

ESA (2007) ENVISAT RA2/MWR Product Handbook. Issue 2.2. https://earth.esa.int/pub/ESA_DOC/ENVISAT/RA2-MWR/ra2-mwr.ProductHandbook.2_2.pdf

Garrett C, Bormans M, Thompson K (1990) Is the exchange through the Strait of Gibraltar maximal or submaximal. In: Pratt LJ (ed) The physical oceanography of sea straits. Kluwer Academic Publishers, Dordrecht

Gómez-Enri J, Cipollini P, Passaro M, Vignudelli S, Tejedor B, Coca J (2016) Coastal altimetry products in the Strait of Gibraltar. IEEE Trans Geosci Remote Sens 54(9):5455–5466. https://doi.org/10.1109/TGRS.2016.2565472

Hughes CW, Bingham RJ, Roussenov V, Williams J, Woodworth PL (2015) The effect of Mediterranean exchange flow on European time mean sea level. Geophys Res Lett 42:466–474. https://doi.org/10.1002/2014GL062654

Izquierdo A, Tejedor L, Sein DV, Backhaus O, Brandt P, Rubino A, Kagan BA (2001) Control variability and internal bore evolution in the Strait of Gibraltar: a 2-D two-layer model study. Estuar Coast Shelf Sci 53(5):637–651. https://doi.org/10.1006/ecss.2000.0706

Lacombe H, Richez C (1982) The regime in the Strait of Gibraltar. In: Nihoul JCJ (ed) Hydrodynamics of semi-enclosed seas, pp. 13–73. Proceedings of the 13th International Liege Colloquium on Ocean Hydrodynamics, Elsevier Oceanography Series. https://doi.org/10.1016/S0422-9894(08)71237-6

Passaro M, Cipollini P, Vignudelli S, Quartly GD, Snaith HN (2014) ALES: A multi-mission adaptive subwaveform retracker for coastal and open ocean altimetry. Remote Sens Environ 145:173–189. https://doi.org/10.1016/j.rse.2014.02.008

Ross T, Garrett C, Le Traon PY (2000) Western Mediterranean sea-level rise: changing exchange flow through the Strait of Gibraltar. Geophys Res Lett 27(18):2949–2952. https://doi.org/10.1029/2000GL011653

Stanichny S, Tigny V, Stanichnaya R, Djenidi S (2005) Wind regime upwelling along the African coast of the Strait of Gibraltar. Geophys Res Lett 32(4). https://doi.org/10.1029/2004GL021760

Absolute Calibration of Sentinel-3A and Jason-3 Altimeters with Sea-Surface and Transponder Techniques in West Crete, Greece

Stelios P. Mertikas, Craig Donlon, Pierre Femenias, Constantin Mavrocordatos, Demitris Galanakis, Thierry Guinle, Francois Boy, Achilles Tripolitsiotis, Xenophon Frantzis, Ilias N. Tziavos, and Georgios S. Vergos

Abstract

This work presents the latest absolute and relative calibrations for the altimeters of the Sentinel-3A and Jason-3 at the Permanent Facility for Altimeter Calibration in west Crete, Greece. Results have been determined at first with the transponder at the CDN1 Cal/Val site on the mountains of west Crete using the ascending Sentinel-3A Pass No. 14 and the descending Jason-3 Pass No.18. Then, sea-surface calibration has been carried out with the descending Sentinel-3A Pass No. 335 and the ascending Jason-3 Pass No.109 based on the Cal/Val facility on Gavdos island. For Sentinel-3A results have been established for cycles 3–27 using Level 2 (Level 0 for transponder) and Non-Time Critical data. For Jason-3, cycles 5–80 and the S-GDR-D data have been worked with the transponder calibration, while for the results with sea-surface calibration, cycles 1–80 with the I-GDR-D data have been implemented.

Sentinel-3A produces biases of the order of a few mm either with the transponder (+2.7 mm) or with the sea-surface calibration (7.3 mm and −4.4 mm). The altimeter of Jason-3 presents a range bias at +22.7 mm (No.18) with the transponder and −36.7 mm for the bias in sea-surface height, respectively. Finally, comparison of sea-surface heights observed by Sentinel-3A relative to Jason-3 demonstrates a difference of +4 cm within a period of ±3 days about 20 km south of Gavdos.

Keywords

Calibration · Jason-3 · Sea-surface · Sentinel-3A · Transponder · Validation

S. P. Mertikas (✉) · X. Frantzis
Geodesy and Geomatics Engineering Lab, Technical University of Crete, Chania, Crete, Greece
e-mail: mertikas@mred.tuc.gr

C. Donlon · C. Mavrocordatos
ESA/ESTEC, Noordwijk, The Netherlands

P. Femenias
ESA/ESRIN, Frascati, Italy

D. Galanakis · A. Tripolitsiotis
Space Geomatica P.C., Chania, Crete, Greece

T. Guinle · F. Boy
Centre National d' Etudes Spatiales (CNES), Toulouse CEDEX, France

I. N. Tziavos · G. S. Vergos
GravLab, Aristotle University of Thessaloniki, Thessaloniki, Greece

1 Introduction

Accuracy, quality and reliability of Earth Observation data and products are strongly dependent upon calibration and validation (Cal/Val) carried out prior and during any satellite operation. At first, individual sensor calibration is performed in specialized labs and under controlled conditions before its satellite launch. Then when in orbit, calibration is mainly accomplished with in-situ observations at dedicated Cal/Val ground facilities. These Cal/Val facilities are committed to monitoring and accessing the performance of satellite instruments and ensure that products from different missions may be merged to deliver long-term and consistent observations of the same environmental parameter.

S. P. Mertikas, R. Pail (eds.), *Fiducial Reference Measurements for Altimetry*,
International Association of Geodesy Symposia 150, https://doi.org/10.1007/1345_2019_63

The Permanent Facility for Altimeter Calibration (PFAC) in west Crete, Greece has been providing Cal/Val services for satellite altimeters since 2004. It employs diverse methodologies that deliver absolute but also relative calibration of past (i.e., Jason-1 & -2, SARAL/AltiKa, Envisat, HY-2A) and current (i.e., Sentinel-3A, Sentinel-3B, CryoSat-2, Jason-3) altimetry missions (Mertikas et al. 2010, 2015; Bonnefond et al. 2012).

In this work, the latest Cal/Val results for ascending and descending passes of the Sentinel-3A and Jason-3 are given. Section 2 presents briefly the infrastructure, instrumentation and setting of the PFAC. Then, calibration results are given with the transponder and the sea-surface techniques in Sects. 3.1 and 3.2. Section 3.3 provides crossover analysis results of Sentinel-3A against Jason-3 at a location about 20 km south of Gavdos. Finally, Sect. 4 summarizes the findings for Sentinel-3A and Jason-3 and introduces plans for the future.

2 The Permanent Facility for Altimeter Calibration

The PFAC is comprised of several Cal/Val sites distributed along the coastlines and the mainland of Crete and Gavdos islands (Fig. 1). In this work, in-situ observations from two of these sites are employed: the CDN1 transponder and the Gavdos coastal Cal/Val sites to determine altimeter calibrations, respectively.

2.1 Transponder Calibration at the CDN1 Cal/Val Site

The CDN1 Cal/Val site lies under a triple crossover location of Sentinel-3A, Sentinel-3B and Jason-3 ground tracks (also underneath of SWOT, SARAL, etc.) on the mountains of Crete at an elevation of 1,100 m. It has been supporting regular calibration with a transponder for S-3A, S-3B, Jason-2 and Jason-3 but also for CryoSat-2 as of October 2015. It is equipped with a Ku-band microwave transponder, two multi-constellation multi-frequency Global Navigation Satellite Systems (GNSS) for absolute positioning and atmospheric delay estimation, and two automated meteorological sensors for measuring humidity, temperature, pressure, wind speed and direction (Fig. 2-left). Hybrid power supply and communications links systems support the continuous and uninterrupted operation of the CDN1 Cal/Val site.

The ascending pass No. 14 of Sentinel-3A and the descending pass No.18 of Jason-3 pass over this CDN1 Cal/Val site. Thus, the transponder receives the signal emitted by the satellite altimeter, amplifies it and transmits it, under controlled and well known conditions, back to the satellite. As the transponder echo is distinguishable on the

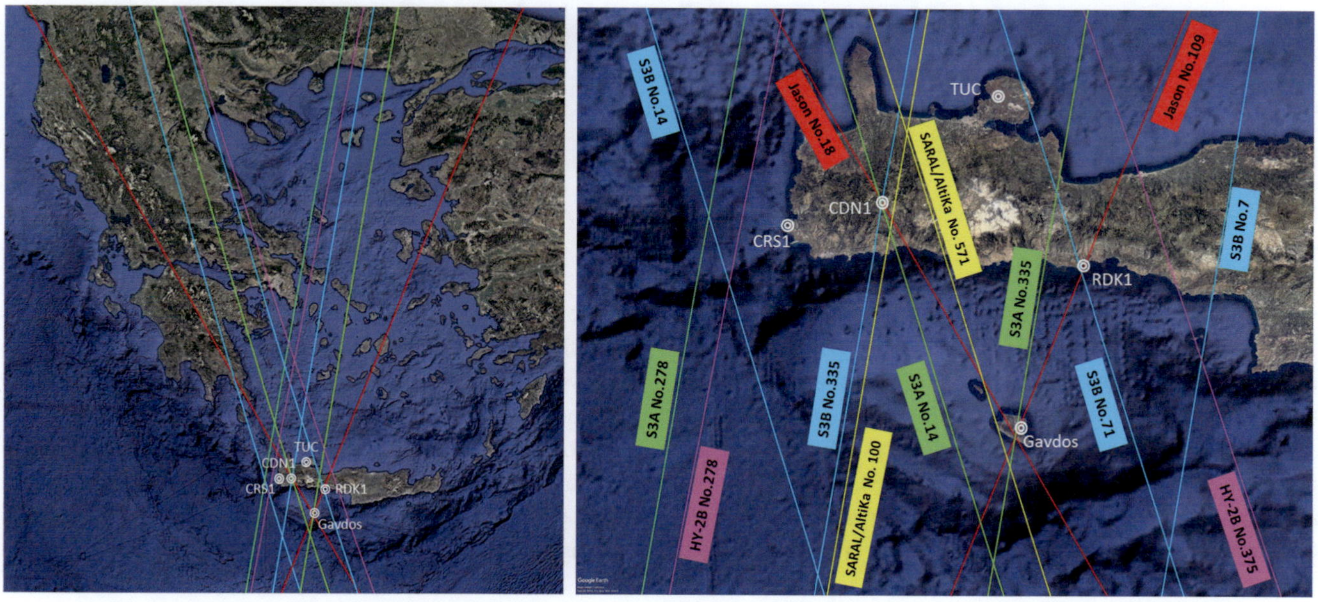

Fig. 1 *Left*: The Permanent Facility for Altimeter Calibration in west Crete, Greece, the independent Cal/Val sites and the ground tracks of currently operating satellite altimeters (Red: Jason-3, Green: Sentinel-3A, Blue: Sentinel-3B, Pink: HY-2A). *Right*: past and current satellite altimetry missions ground tracks over west Crete, Greece

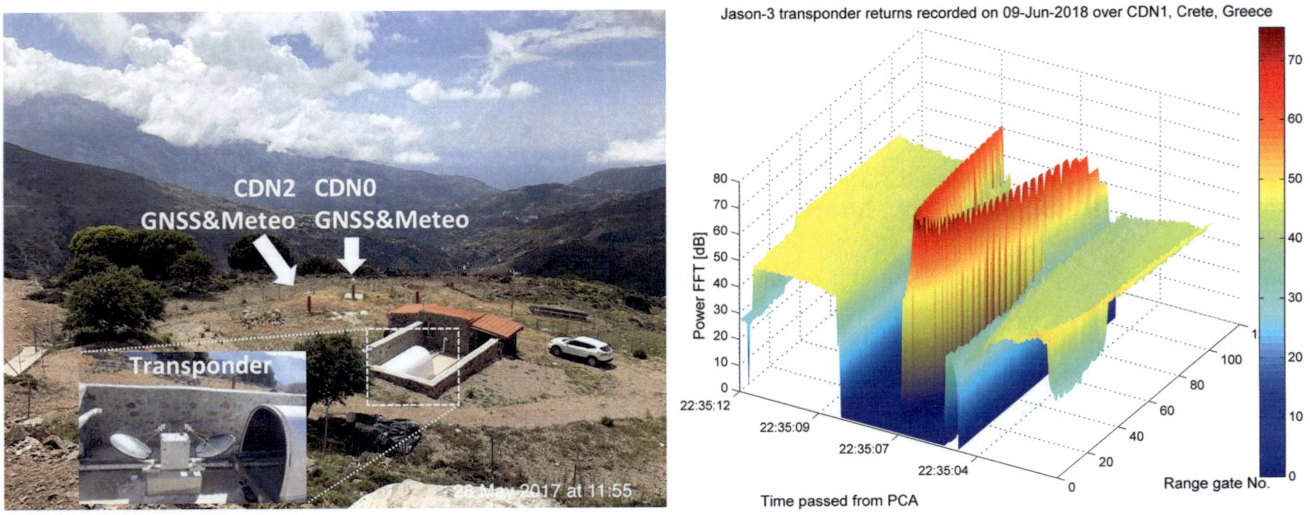

Fig. 2 *Left*: An overview of the CDN1 Cal/Val transponder site on west Crete mountains, Greece. *Right*: Transponder echo returns of Jason-3, as measured on 9-June-2018 (Pass No.18, Cycle 86)

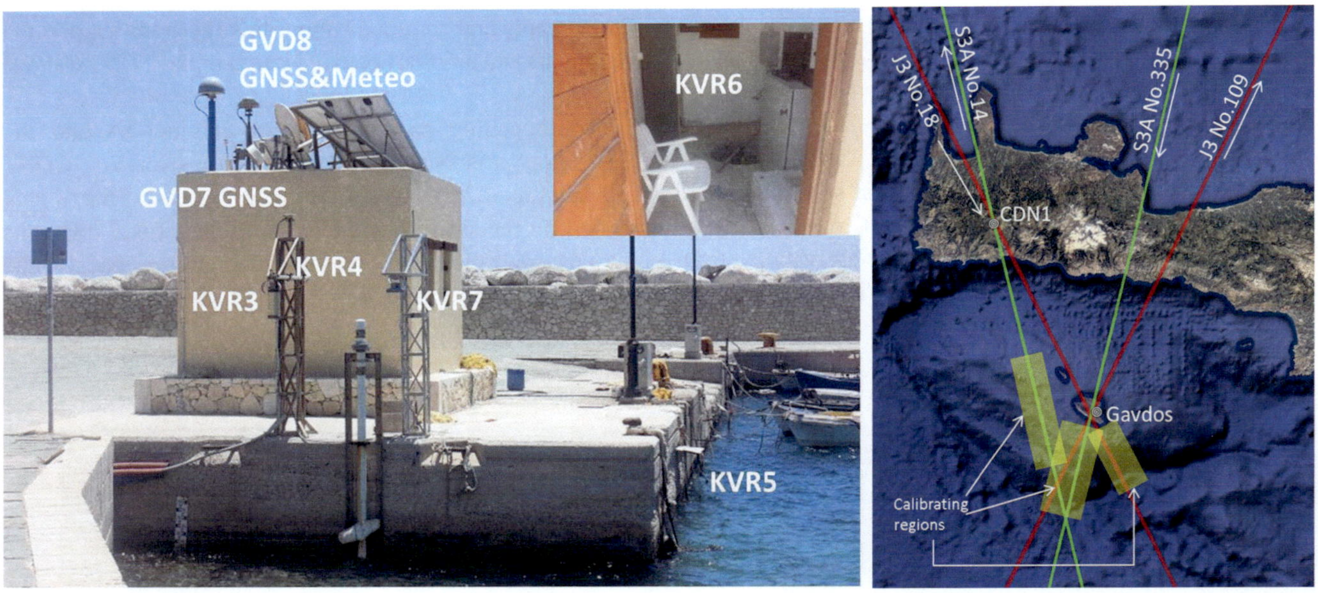

Fig. 3 *Left*: The "Karave" sea-surface Cal/Val site instrumentation in Gavdos harbor. *Right*: The calibrating regions in the open sea for Sentinel-3A and Jason-3 altimeter missions

satellite records (Fig. 2-right), it can be clearly isolated and further processed. In that manner, the range between the satellite and the transponder can be independently deduced (Mertikas et al. 2018).

Calibration results are directly influenced by the wet troposphere delay of satellite signals. This has to be estimated as the satellite radiometer does not regularly operate over the transponder land location. Wet delays are at present derived by GNSS processing. In the future, delays will be determined by radiometers. The standard uncertainty in estimating the wet troposphere delay as derived by GNSS, is of the order of ±8 mm (Mertikas et al. 2018). Other critical parameters in transponder's calibration are the internal path delay of its electronics, its electronic phase center, their variations and stabilities, etc. As a result of this process, the transponder can provide absolute calibration by observing the range when the satellite flies over this site for about 3–5 s.

2.2 Sea-Surface Calibration at the Gavdos Cal/Val Facility

The Gavdos Cal/Val facility has been providing altimeter biases since 2004 with sea-surface calibration. Several tide gauges of diverse measuring principles (i.e., radar, pressure, acoustic) determine the water level at the Gavdos island harbor. These observations are transformed into ellipsoidal heights with collocated GNSS and meteorological sensor data (Fig. 3). As altimeter measurements are contaminated by the land of Gavdos, sea-surface calibration has to take place in open sea. Thus, Cal/Val reference data determined at the ground-truth facility on the island need to be transferred over a region where altimeters measure uncontaminated (Fig. 3-right). After various investigations (Mertikas et al. 2010), this region has been selected to be 10–15 km south of the Gavdos coastline.

3 Calibration Results

The latest absolute and relative calibration results for Sentinel-3A and Jason-3 altimeters are given in the following sections.

3.1 Sea-Surface and Transponder Results for Sentinel-3A

Sentinel-3A, on its ascending Pass No.14, starts off from south, moves west of Gavdos, where sea-surface calibration takes place, and then a couple of seconds later it reaches the CDN1 Cal/Val site on the mountains for a consecutive transponder calibration. Sentinel-3A biases are presented in Fig. 4. Sea-surface calibrations are given for cycles 7–27, using marine, Non-Time-Critical, SAR-mode and Level-2 products. Transponder results are based upon Level-0 and Level-1A SAR data and correspond to cycles 3–22. Another sea-surface calibration follows for the descending pass No. 335 of S3A for cycles 5–26 and Level-2 products. All these Cal/Val results for Sentinel-3A with sea-surface and transponder techniques are given on Fig. 4.

3.2 Sea-Surface and Transponder Results for Jason-3

Jason-3, on its descending Pass No.18, sets out from north over the transponder, continues on its south-east orbit and passes over an open sea area, west of the Gavdos Cal/Val facility. Thus, the same satellite pass is calibrated using both sea-surface (Fig. 5a) and transponder (Fig. 5c) techniques with Interim GDR-D for cycles 1–86 and S-GDR-D products for cycles 5–80, respectively. The ascending Pass No.109 is also calibrated for its cycles 1–80 (Fig. 5b).

3.3 Crossover Calibrations of Sentinel-3A and Jason-3

The crossover calibration involves the direct comparison of sea-surface heights of Sentinel-3A Pass No.18 and Jason-3 Pass No.109 when they overpass the same location (about 20 km south of Gavdos) within a period of ±3 days. Results are shown in Fig. 6.

Table 1 sums up these calibration results for Sentinel-3A and Jason-3 as obtained at the PFAC using diverse and independent methodologies.

4 Conclusions

This work has presented the latest results for Sentinel-3A and Jason-3 altimeters with transponder, sea-surface and crossover calibrations for ascending and descending passes. A list of short statements of the main points could be as follows: (1) The altimeter bias of S-3A is less than ±1 cm irrespective of the calibration method and the Cal/Val site employed; (2) No directional errors are monitored for S-3A, whereas some directional disparity of 2–3 cm appears to exist for Jason-3; (3) Sentinel-3A seems to measure sea surface higher than Jason-3; (4) The repeat pattern of each satellite orbit (27-day of S-3A versus 10-day of Jason-3) may produce irregularities in altimeter bias comparisons; and (5) Transponder and sea-surface calibrations deliver results of the same level of magnitude for S3A and Jason-3.

The Cal/Val results in this work for Sentinel-3A and Jason-3 are consistent with the ones presented by other permanent Cal/Val sites over the globe (OSTST 2018). The maximum deviation between independent Cal/Val results lie within the range of ±2 cm at most. At present, this provides evidence that Sentinel-3A and Jason-3 meet their target specification of ±3 cm for sea-surface height measurements.

At the moment, the PFAC is in the process of upscaling its operational capabilities by (1) embracing and adopting the concept of Fiducial Reference Measurement (Donlon 2018) for altimetry (2) construction of new Cal/Val sites and improving its instrumentation.

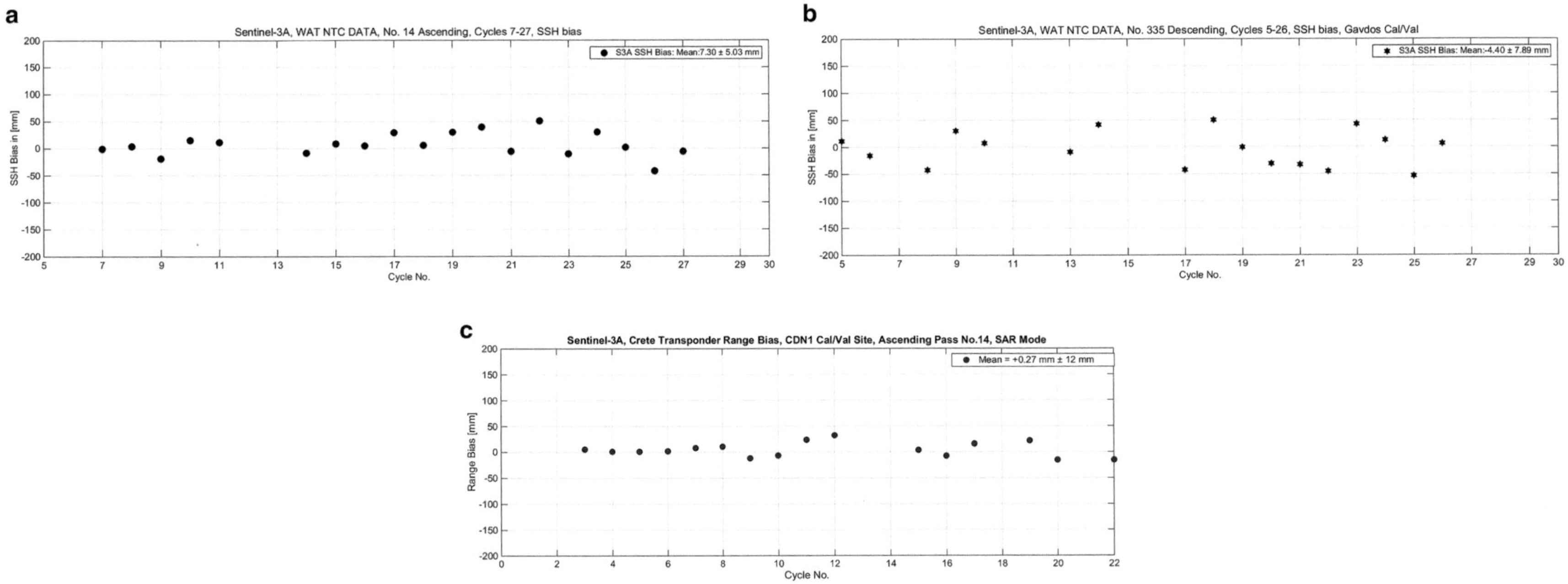

Fig. 4 Sea-surface calibrations for S-3A Pass No.14 (**a**) and Pass No.335 (**b**). Transponder calibration results for S-3A Pass No.14 (**c**)

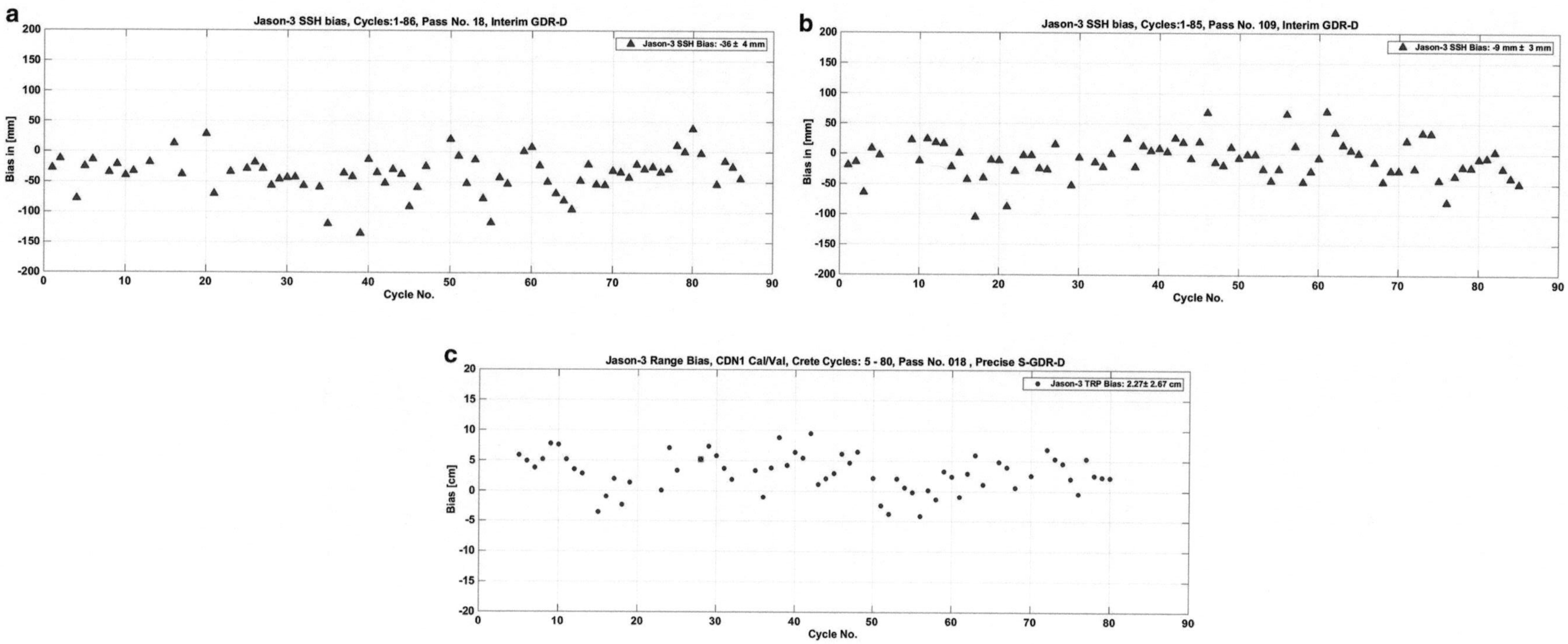

Fig. 5 Sea-surface calibration results for Jason-3 Pass No.18 (**a**) and Pass No.109 (**b**). Transponder calibration results for Jason-3 Pass No.18 (**c**)

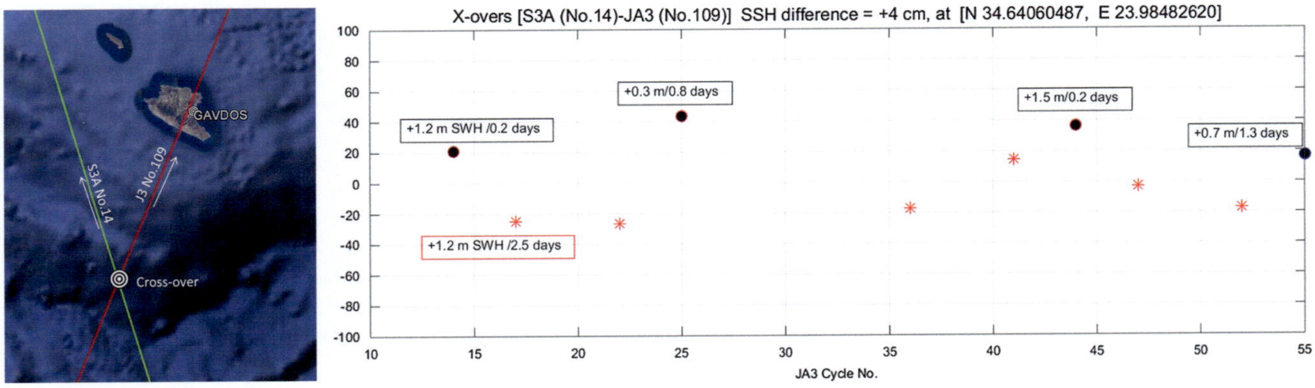

Fig. 6 The S-3A Pass No.14 and Jason-3 Pass No.109 crossover location south of Gavdos (left) and the differences in sea-surface heights with respect to their temporal offset and SWH (right)

Table 1 Absolute direct (transponder), absolute indirect (sea-surface) and relative (crossover) calibration results for Sentinel-3A and Jason-3 altimeters missions at the PFAC in west Crete, Greece

Satellite	Pass No.	Cycles	Product	Method	Bias [mm]
S-3A	14	3–22	WAT, NTC, L1A, L0	Transponder	+2.7 mm
S-3A	14	7–27	WAT, NTC, L2	Sea-surface	+7.3 mm
S-3A	335	5–26	WAT, NTC, L2	Sea-surface	−4.4 mm
Jason-3	18	5–80	S-GDR-D	Transponder	+22.7 mm
Jason-3	18	1–80	I-GDR-D	Sea-surface	−36.7 mm
Jason-3	109	1–80	I-GDR-D	Sea-surface	−9.1 mm
S3A–J3	14–109	13 cases		Cross-over	+40.0 mm

Acknowledgements This work has been supported and funded by the EU and ESA (ESA Contracts No. 4000117101/16/I/BG and 400122240/17/I-BG). The operational work for Jason-3 transponder Cal/Val has been supported by CNES.

References

Bonnefond P, Desjonqueres J-D, Haines B, Mertikas S, Watson C (2012) Absolute calibration of the Topex/Poseidon and Jason measurement systems: twenty years of monitoring from dedicated sites. In: Proceedings of the ESA Symposium "20 Years of progress in Radar Altimetry", Venice Lido, Italy

Donlon C (2018) Fiducial reference measurements for altimetry. FRM4ALT Project Webportal. https://goo.gl/Yn23pQ. Accessed 26 June 2018

Mertikas S, Ioannides RT, Tziavos IN et al (2010) Statistical models and latest results in the determination of the absolute bias for the radar altimeter of Jason satellites using the Gavdos facility. Mar Geod 33(1):114–149. https://doi.org/10.1080/01490419.2010.488973

Mertikas SP, Zhou X, Qiao F, Daskalakis A, Lin M, Peng H, Tziavos IN, Vergos G, Tripolitsiotis A, Frantzis X (2015) First preliminary results for the absolute calibration of the Chinese HY-2 altimetric mission using the CRS1 calibration facilities in West Crete, Greece. Adv Space Res 57(1):78–95. https://doi.org/10.1016/j.asr.2015.10.016

Mertikas S, Donlon C, Féménias P, Mavrocordatos C, Galanakis D, Tripolitsiotis A, Frantzis X, Kokolakis C, Tziavos IN, Vergos G, Guinle T (2018) Absolute Calibration of the European Sentinel-3A Surface Topography mission over the permanent facility for altimetry calibration in West Crete, Greece. Remote Sens 10(11):1808. https://doi.org/10.3390/rs10111808

OSTST (2018) In: Bonnefond P, Willis J, Leuliette E, Scharroo R, Donlon C (eds) Report of the 2018 ocean surface topography science team meeting. Available online: https://www.aviso.altimetry.fr/fileadmin/documents/OSTST/2018/OSTST_2018_Meeting_Report_Final.pdf. Accessed 15 Feb 2019

Multi-Mission Cross-Calibration of Satellite Altimeters

Systematic Differences Between Sentinel-3A and Jason-3

Denise Dettmering and Christian Schwatke

Abstract

The inter-mission cross-calibration is a basic prerequisite for long-term sea level change studies on all spatial scales. Especially, for climate studies the consistent combination of successive missions is essential. This study uses a global multi-mission crossover analysis in order to investigate the performance of the Copernicus Sentinel-3A altimetry mission and its consistency to consecutive missions such as Jason-3. The first 1.5 years of data show an inter-mission bias of 3.6 cm with respect to Jason-3 with a linear trend of 4.0 mm/year. When using Pseudo Low Resolution Mode (PLRM) instead of Delay-Doppler data the bias increases whereas the trend decreases. Given the short time period under investigation these numbers should not be overrated, however, a careful future monitoring is necessary.

Keywords

Cross-calibration · GCE · Range bias drift · Sentinel-3A

1 Introduction

Most sea level change studies, especially when focused on climate research, require reliable long-term observation time series. However, the majority of satellite missions are only available for a couple of years and, usually, follow-on missions provide data sets not fully consistent with the original mission. Moreover, it might be necessary to combine contemporaneous missions in order to improve data resolution and accuracy. Hence, to ensure a consistent long-term data set with optimal spatio-temporal resolution a careful calibration is necessary.

The relative inter-mission cross-calibration is one of the basic methods of calibration/validation (CAL/VAL) activities (Dorandeu et al. 2004) and an inevitable prerequisite for all applications based on multi-mission altimetry. In contrast to in-situ calibrations (e.g. Haines et al. 2010; Watson et al. 2011; Mertikas et al. 2016) a crossover analysis provides relative, but global calibration results and allows for estimating geographically distributed error pattern. Furthermore, since this method does not directly need simultaneous measurements at nearly the same position, the approach is independent of the satellites orbit configuration and can also be used after the usual tandem phase of follow-on missions (e.g. as for Jason-2 and Jason-3 (Boy 2016)). As a consequence, the calibration can be performed as continuous cross-calibration over the whole mission lifetime. In contrast to calibration approaches using tide gauge measurements (Mitchum 2000) no external information is necessary. Moreover, no addition uncertainties due to these external data have to be taken into account. However, this comes at the cost that only inter-mission relative differences can be detected without knowing the absolute errors of the single instruments.

DGFI-TUM is performing multi-mission altimeter crossover analysis (MMXO) on a regular basis in order to estimate relative radial errors between the different altimeter systems operating simultaneously. Here, results for the Sentinel-3A mission with respect to Jason-3 will be presented. First, the method of MMXO is shortly recapped, and the data sets used within the analysis are introduced.

D. Dettmering (✉) · C. Schwatke
Deutsches Geodätisches Forschungsinstitut der Technischen Universität München (DGFI-TUM), München, Germany
e-mail: denise.dettmering@tum.de

S. P. Mertikas, R. Pail (eds.), *Fiducial Reference Measurements for Altimetry*,
International Association of Geodesy Symposia 150, https://doi.org/10.1007/1345_2019_58

Section 3 is focused on the calibration results (inter-mission offset, range bias drift, geographically pattern) and a discussion. The paper finishes with conclusion and outlook.

2 Multi-Mission Cross-Calibration

The cross-calibration is performed between all contemporaneous altimeter systems. It is based on sea surface height (SSH) differences at crossovers between ascending and descending passes of single missions as well as dual-satellite crossovers. These differences are minimized within a least squares adjustment process, and for each measurement a radial error is estimated. In order to ensure a certain smoothness of the along-track errors of a single mission (without introducing an analytic function) also consecutive SSH difference are minimized. The total set of crossover differences creates a highly redundant network and enables a stable estimate of radial errors with a dense sampling for all altimeter systems analyzed. An iterative variance component estimation is applied to obtain an objective relative weighting between the different altimeter systems. Moreover, a latitude dependent weighting is introduced to account for increasing numbers of crossover locations towards polar areas. When computing global averages from the estimated radial errors (see Sect. 3) the in-homogeneous geographical distribution of crossover points may slightly influence the results. However, since this effect is very similar for different epochs, the relative values as well as the temporal evolution (i.e. drift behavior) will not be impacted. The approach is described in detail in Bosch (2007) and Bosch et al. (2014) and is already successfully applied for the calibration of various missions, such as Jason-2 (Dettmering et al. 2010) and Saral (Dettmering et al. 2015) as well as for analyzing the impact of orbit determination parameters on inter-mission consistency (e.g. Rudenko et al. 2014, 2018).

Sentinel-3A was launched in February 2016. It provides a lot of different data sets, including some reprocessing. Here, None Time Critical (NTC) Level 2 marine data from processing baseline PB 2.27 are used (reprocessing REP_006_SM2_W_new). The observations cover a time period of about 1.5 years between March 1st, 2016 and December 15th, 2017. Sentinel-3A is measuring in Synthetic Aparture Radar (SAR) mode globally. However, in addition, Pseudo Low Resolution Mode (PLRM) data are provided, i.e. the SAR pulses are used to compute classical pulse-limited waveforms. This product is also used here for some specific tests. Most of the original geophysical corrections included in the NTC data set are applied to the data. However, some corrections are replaced by external products in order to ensure the best consistency with the other missions involved in the processing, for which the same correction models are applied.

In addition to Sentinel-3 the following missions are used in this study: Jason-3, Jason-2 (on core orbit and interleaved orbit), and SARAL (most of the time in drifting phase). For the atmospheric corrections identical models are used for all missions, i.e. NIC09 (Scharroo and Smith 2010) for the ionosphere and ECMWF VMF1 (Boehm et al. 2009) for wet and dry troposphere. This allows to assign the detected systematic differences directly to the altimeter instruments (or its orbits) and exclude any impact from other instruments such as the microwave radiometer. For the same reason external models for ocean tides (EOT11a (Savcenko and Bosch 2008)) and dynamic atmospheric correction (DAC[1]) are applied to all missions involved in the study. The only correction that differs between the different missions is the sea state bias (SSB) correction. Here, the corrections included in the original data sets are used. For Sentinel-3A two different SSB corrections are available, one for SAR and one for PLRM data. However, they both are based on empirical model solutions fitted on Jason-2 GDR-C data, which also holds for the SSB correction included in the Jason-3 data set. This might be critical since it is well known that for SAR altimeters the wave heights have a different impact on SSB as for conventional altimeters (Passaro et al. 2016).

3 Results and Discussion

Radial Errors and Its Decomposition The main result of the MMXO are time series of radial errors for each mission involved in the process. These are estimated at the crossover points and interpolated to every single measurement. The basic objective of these values is the application as range correction in order to create a consistent multi-mission measurement data set. In addition, the post-processing analysis of these time series allows for the detection and analysis of systematic differences between the missions (such as instrument drifts or discrepancies in orbit computation).

In the post-processing, the radial errors x_i are decomposed into a range bias Δr and center-of-origin shifts $\Delta x, \Delta y, \Delta z$ following Eq. (1) where ϕ_i, λ_i are the crossover locations defined on the TOPEX-Ellipsoid and the reference frame is ITRF2008, as defined by the orbits of the used satellites. This equation is solved per cycle (i.e. about 10

[1]Dynamic Atmospheric Corrections are produced by CLS Space Oceanography Division using the Mog2D model from Legos and distributed by Aviso, with support from CNES (http://www.aviso.altimetry.fr/).

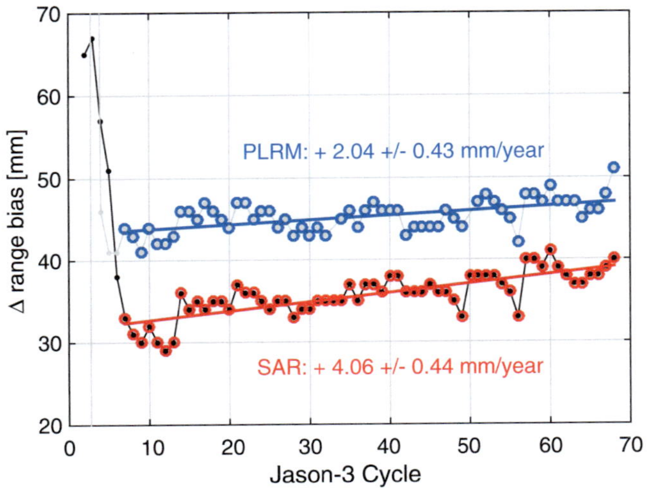

Fig. 1 Sentinel-3A SAR (red) and PLRM (blue) relative range bias Δr w.r.t. Jason-3

days) using a least squares adjustment with errors e_i.

$$x_i + e_{x_i} = \Delta r + \Delta x \cdot cos(\phi_i)cos(\lambda_i) + \Delta y \cdot cos(\phi_i)sin(\lambda_i) + \Delta z \cdot sin(\phi_i) \quad (1)$$

Range Bias The estimated range biases for Sentinel-3A relative to Jason-3 are plotted in Fig. 1. For the first five Jason-3 cycles strong systematic effects are visible, which are mainly caused by very few valid Sentinel-3A observations in the very first mission phase. These cycles will be excluded from all further analysis. The remaining time series of 62 cycles shows a mean offset between Sentinel-3A and Jason-3 of 35.7 mm for SAR and 45.3 mm for PLRM. This relative difference of about 1 cm is well in accordance to in situ absolute calibration results in Australia (Watson et al. 2018). In addition to the mean offsets, a systematic temporal behavior is detectable between May 2016 and December 2017 showing a small positive trend of 4 mm/year between Sentinel-3A and Jason-3. Despite the short period of only about 1.5 years, this trend is mathematically significant with an uncertainty of 0.4 mm. Nevertheless, it may also be a non-linear (e.g. long-period) effect. Moreover, the systematic may also be caused by rapid bias changes, e.g. between cycle 13 and 14 (June 2016) and/or 56 and 57 (August 2017). This cannot be clarified before a longer time series is available.

Since the MMXO is only providing relative information, it is not clear which of the missions (Sentinel-3A or Jason-3 or both) is causing the drift behavior. Currently, from the existing in-situ calibrations, no proof for a drift in Sentinel-3A data (Bonnefond et al. 2018; Watson et al. 2018) nor in Jason-3 data (OSTST 2017; Watson et al. 2018) is available. However, ESA (2018) provides an absolute range bias of

Table 1 Sentinel-3A range bias (SAR and PLRM) w.r.t. three other missions

S3A mode	Mission	Bias [mm]	Trend [mm/year]	Cycles #
SAR	Jason-3	35.7 ± 2.6	4.0 ± 0.4	7–68
PLRM	Jason-3	45.3 ± 1.9	2.0 ± 0.4	7–68
SAR	Jason-3	34.6 ± 2.2	5.3 ± 0.8	7–46
SAR	Jason-2	11.8 ± 2.3	4.3 ± 1.0	7–46
SAR	SARAL	64.0 ± 2.7	5.0 ± 1.1	7–46
PLRM	Jason-3	44.6 ± 1.5	0.9 ± 0.8	7–46
PLRM	Jason-2	21.9 ± 1.8	0.3 ± 0.9	7–46
PLRM	SARAL	74.5 ± 2.2	1.1 ± 1.1	7–46

The values w.r.t. Jason-3 are provided for two different time series lengths

8 mm and a drift of 3.2 mm/year (computed from cycles 3–28 with different processing baselines). Even if this value is not significant given the uncertainty of the linear fit, it might be an indication that the Sentinel-3A range bias is not stable.

As the MMXO also provides range bias information for the other missions involved, these can be checked too. Table 1 summarizes the trend estimates of Sentinel-3A with respect to all other missions. In order to provide consistent results, the analysis period has been restricted until the end of Jason-2 (interleaved orbit). From these values it is obvious that the drift originates from the Sentinel-3A data set since it is visible with respect to all three other missions in the same order of magnitude (taking the large uncertainties induced by the short computation period of only about 1 year into account).

The source for this trend can be related to a drift in the altimeter instrument, but also in systematic effects coming from one of the used corrections (that is not consistent with the other missions). One likely cause is the SSB correction that is not yet tuned for Sentinel-3A (EUMETSAT 2018). Even if it is known that the interaction of SAR measurements with waves differ from that of LRM measurements, the same empirical correction model is used to generate the SBB corrections, since no dedicated Sentinel-3A model is available until now. A more detailed investigation on this is for sure necessary. Moreover, at least part of the effect can be explained by a known Sentinel-3A SAR SSB drift resulting from a drift in the point target response (PTR) width (Dinardo et al. 2018), which leads to a drift in significant wave heights differences between SAR and PLRM.

Interestingly, the trend reduces from 4 to 2 mm/year when using PLRM data (range and SSB correction) instead of SAR data. When limiting the period to cycle 7 to 46 (about 1 year) no trend in the PLRM data is detectable anymore (see Table 1 and Fig. 1).

It is important to keep in mind, that the drift information should be interpreted tentatively since a 1.5 year period is certainly not long enough for providing reliable numbers.

Table 2 Sentinel-3A center-of-origin w.r.t. Jason-3 (cycles 7–68)

S3A mode	Component	Bias [mm]	Trend [mm/year]
SAR	Δx	0.8 ± 1.8	-1.6 ± 0.4
SAR	Δy	-1.4 ± 1.9	0.4 ± 0.5
SAR	Δz	-2.3 ± 2.6	-0.8 ± 0.7
PLRM	Δx	0.7 ± 1.3	-1.0 ± 0.3
PLRM	Δy	-1.3 ± 1.7	0.4 ± 0.4
PLRM	Δz	-0.1 ± 2.9	-1.4 ± 0.7

The upper part is computed based on Sentinel-3A SAR measurements, the lower part on PLRM measurements

Instead, they should be taken as an indication, that some systematic effects are inherent in the data set, which should be pursued in the future.

Center-of-Origin Realization In addition to the range bias, the decomposition of the time series of radial errors also provides information about the different realization of the center-of-origin by the different missions (see Eq. (1)). Here, no systematic differences can be seen when comparing Sentinel-3A to Jason3 as visible from Table 2. Even if a slight negative trend in Δx seems to be visible as well as a negative offset in Δy and Δz, most of these are not mathematically significant. Except the drift in Δz, all these values reduces when using PLRM data. When a longer time series is available this behavior should be studied in more detail. Those effects can be related to systematic differences in the orbit computation of different missions. They will reflect as geographically correlated differences, an error component investigated in more detail in the following.

Geographically Correlated Errors For regional sea level studies, the knowledge of geographically correlated error (GCE) components is critical. GCE arise from common components in the errors of ascending and descending passes and map directly in the derived sea surface heights. Even if they are not detectable through single-satellite crossover analysis, they can be estimated based on MMXO-derived radial errors Δr following the theory of Rosborough (1986). For their computation, the radial errors of ascending Δr^{asc} and descending passes Δr^{desc} are averaged separately in predefined geographic grid cells (the mean values are indicated by an overbar in Eq. (2)) and used to generate a mean $\Delta\gamma$ and a variable part $\Delta\delta$ per grid cell (Eq. (2)). The mean error component directly provides the GCE.

$$\begin{aligned} \Delta\gamma &= (\overline{\Delta r^{asc}} + \overline{\Delta r^{desc}})/2 \\ \Delta\delta &= (\overline{\Delta r^{asc}} - \overline{\Delta r^{desc}})/2 \end{aligned} \tag{2}$$

The GCE of all missions, as plotted in Fig. 2, remain well below the 2 cm. The largest values can be seen for SARAL in the area north of Australia. For Sentinel-3A, 95% of the values are smaller than 4.5 mm with the biggest effects in the area of the Indian Ocean. In comparison with Jason-3 (3.5 mm) and Jason-2 (3.8 mm) this number is slightly higher but still almost half of the value for SARAL (7.4 mm). Polar areas with latitudes above 60° are excluded from this analysis in order to allow for a fair inter-mission comparison. Most of these effects will probably be caused by precise orbit determination, which seems to be consistent for all missions.

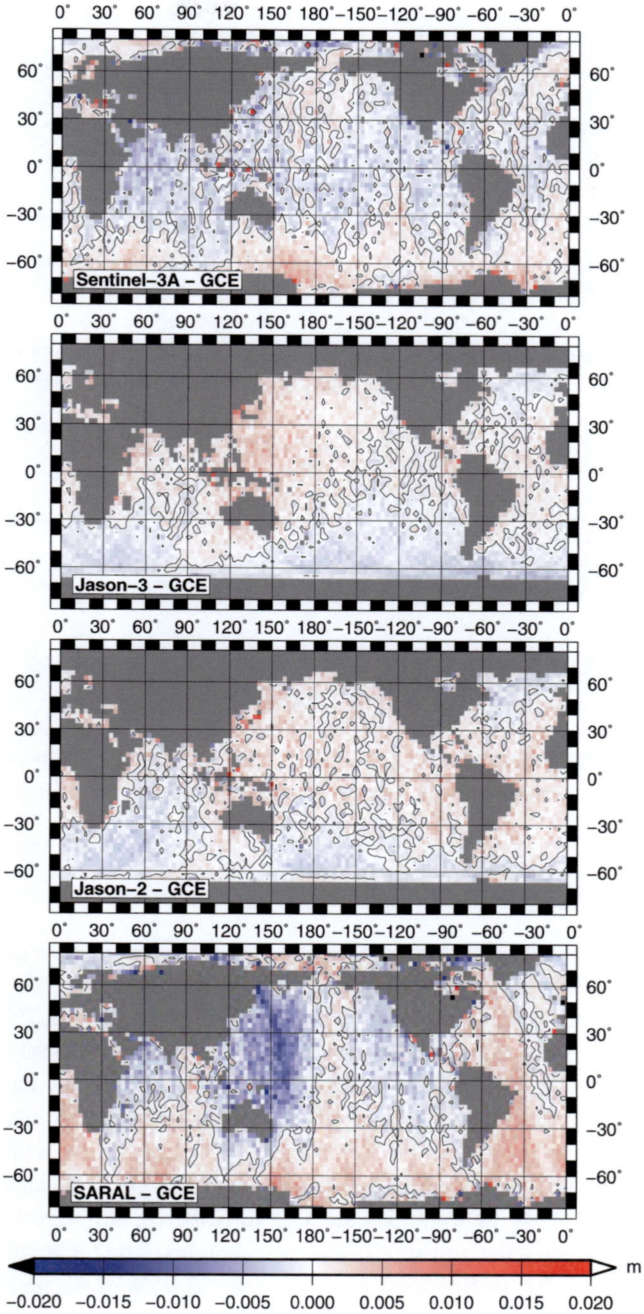

Fig. 2 Geographically correlated errors for Sentinel-3A (top) and the other three missions included in the MMXO analysis

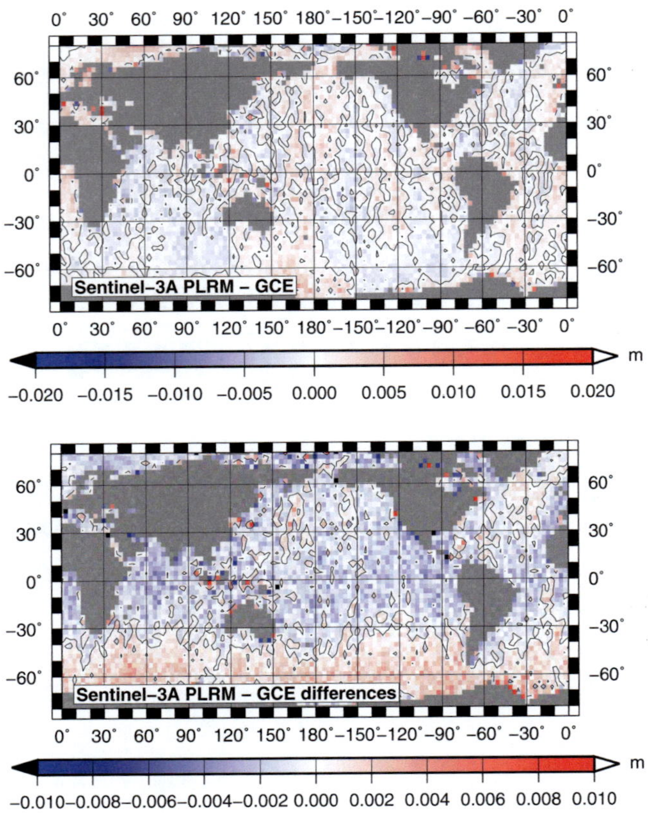

Fig. 3 Geographically correlated errors for Sentinel-3A PLRM (top) and SAR-PLRM difference (bottom)

When analyzing the Sentinel-3A PLRM data instead of the SAR data, the GCE per grid cell are changing. The differences remain smaller than 5 mm for most areas (98.7% of the grid cells). However, they show a clear north-south pattern as illustrated in the lower plot of Fig. 3. Since the orbit is not changed between the versions, this effect is only related to range and SSB differences. The correlation with the geographical distribution of mean wave heights indicates that most of these effects are due to SSB, i.e. differences in altimetry derived significant wave heights. The GCE for Sentinel-3A PLRM stay below 3.8 mm for 95% of the area (see Fig. 3, top plot).

4 Conclusion and Outlook

In this study, a multi-mission crossover analysis is used in order to calibrate Sentinel-3A with respect to Jason-3 and two other missions. The applied method provides time series of radial errors for the whole period under investigation, which are then used to derive inter-mission range biases and their drift behavior, as well as information on geographically correlated (mean) errors and the center-of-origin realization of each mission included in the analysis.

The analysis shows a Sentinel-3A SAR range bias of 3.6 cm with respect to Jason-3 with a significant drift behavior of 4.0 mm/year, which is believed to originate from Sentinel-3A since it is also detectable with respect to Jason-2 and SARAL. When using PLRM data the bias increased to 4.5 cm, however, the trend reduces to 2.0 mm/year. These values are based on model corrections for atmospheric attenuation, since the radiometer correction for Sentinel-3A is not yet fully tuned (EUMETSAT 2018). The investigation shows no systematic effects coming from orbit computation since center-of-origin realization as well as geographically correlated error pattern are consistent to other missions. However, when analyzing the GCE differences between SAR and PLRM solutions of Sentinel-3A, a small but obvious north-south pattern is detected, probably related to different performances of SSB corrections for SAR and LRM.

The time period under investigation only comprises about 1.5 years of data. Therefore, the trend estimation is still unreliable and should only be seen as a first indication that systematic inter-mission effects are present between Sentinel-3A and Jason-3. A continuous monitoring of these inconsistencies are essential for the future, especially taking into account that the estimated trend is in the same order of magnitude than current estimates for global mean sea level rise.

Acknowledgements We kindly thank the institutions in charge of mission operation and maintenance (ESA, EUMETSAT, NASA, CNES) as well as the data providers (EUMETSAT, AVISO). Special thank goes to EUMETSAT for supporting the participation at the International Review Workshop On Satellite Altimetry Cal/Val Activities and Applications, Chania, Greece, April 2018.

References

Boehm J, Kouba J, Schuh H (2009) Forecast vienna mapping functions 1 for real-time analysis of space geodetic observations. J Geodesy 83(5):397–401. https://doi.org/10.1007/s00190-008-0216-y

Bonnefond P, Laurain O, Exertier P, Boy F, Guinle T, Picot N, Labroue S, Raynal M, Donlon C, Féménias P, Parrinello T, Dinardo S (2018) Calibrating the SAR SSH of Sentinel-3A and CryoSat-2 over the corsica facilities. Remote Sensing 10(1):92. https://doi.org/10.3390/rs10010092

Bosch W (2007) Discrete crossover analysis. In: Tregoning P, Rizos C (eds) Dynamic planet - monitoring and understanding a dynamic planet with geodetic and oceanographic tools, IAG Symposium, vol 130. Springer, Berlin

Bosch W, Dettmering D, Schwatke C (2014) Multi-mission cross-calibration of satellite altimeters: constructing a long-term data record for global and regional sea level change studies. Remote Sensing 6(3):2255–2281. https://doi.org/10.3390/rs6032255

Boy F (2016) Continuity of the altimetry constellation: JASON-3 first year of observations over ocean. In: AGU Fall Meeting Abstracts G21A-0977

Dettmering D, Bosch W (2010) Global calibration of jason-2 by multi-mission crossover analysis. Marine Geodesy https://doi.org/10.1080/01490419.2010.487779

Dettmering D, Schwatke C, Bosch W (2015) Global calibration of saral/altika using multi-mission sea surface height crossovers. Mar Geodesy 38(suppl 1):206–218. https://doi.org/10.1080/01490419.2014.988832

Dinardo S, Scharroo R, Lucas B, Martin-Puig C, Nogueira-Loddo C (2018) Impact of the Sentinel-3A SRAL PTR width drift on the L2 marine measurement. Poster presented at OSTST 2018, Ponta Delgada, Azores

Dorandeu J, Ablain M, Faugère Y, Mertz F, Soussi B, Vincent P (2004) Jason-1 global statistical evaluation and performance assessment: calibration and cross-calibration results. Mar Geodesy 27(3–4):345–372. https://doi.org/10.1080/01490410490889094

ESA Sentinel Online, Technical Guides. https://sentinels.copernicus.eu/web/sentinel/technical-guides/sentinel-3-altimetry/cal-val/calibration

EUMETSAT Sentinel-3A Product Notice - STM L2 Marine. Product Notice ID: EUM/OPS-SEN3/DOC/16/893228

Haines BJ, Desai SD, Born GH (2010) The harvest experiment: calibration of the climate data record from topex/poseidon, jason-1 and the ocean surface topography mission. Mar Geodesy **33**(suppl 1):91–113. https://doi.org/10.1080/01490419.2010.491028

Mertikas SP, Zhou X, Qiao F, Daskalakis A, Lin M, Peng H, Tziavos IN, Vergos G, Tripolitsiotis A, Frantzis X (2016) First preliminary results for the absolute calibration of the Chinese HY-2 altimetric mission using the CRS1 calibration facilities in West Crete, Greece. Adv Space Res 57(1):78–95. https://doi.org/10.1016/j.asr.2015.10.016

Mitchum GT (2000) An improved calibration of satellite altimetric heights using tide gauge sea levels with adjustment for land motion. Mar Geodesy 23:145. https://doi.org/10.1080/01490410050128591

OSTST (2017) Report of the Ocean Surface Topography Science Team Meeting, Miami, FL, USA, October 23–27, 2017. Tech. rep., OSTST. https://www.aviso.altimetry.fr/fileadmin/documents/OSTST/2017/OSTST_2017_Meeting_Report_Final.pdf

Passaro M, Dinardo S, Quartly GD, Snaith HM, Benveniste J, Cipollini P, Lucas B (2016) Cross-calibrating ALES Envisat and CryoSat-2 Delay-Doppler: a coastal altimetry study in the Indonesian Seas. Adv Space Res 58(3):289–303. https://doi.org/10.1016/j.asr.2016.04.011

Rosborough G (1986) Satellite Orbit perturbations due to the Geopotential. Report csr-86-1, Center of Space Research, University of Texas, Austin

Rudenko S, Dettmering D, Esselborn S, Schöne T, Förste C, Lemoine JM, Ablain M, Alexandre D, Neumayer KH (2014) Influence of time variable geopotential models on precise orbits of altimetry satellites, global and regional mean sea level trends. Adv Space Res 54(1):92–118. https://doi.org/10.1016/j.asr.2014.03.010

Rudenko S, Bloßfeld M, Müller H, Dettmering D, Angermann D, Seitz M (2018) Evaluation of DTRF2014, ITRF2014, and JTRF2014 by precise orbit determination of SLR satellites. IEEE Trans Geosci Remote Sensing 56(6):3148–3158. https://doi.org/10.1109/TGRS.2018.2793358

Savcenko R, Bosch W (2008) EOT08a - empirical ocean tide model from multi-mission satellite altimetry. Report no. 81, Deutsches Geodätisches Forschungsinstitut, München

Scharroo R, Smith WHF (2010) A global positioning system-based climatology for the total electron content in the ionosphere. J Geophys Res 115(A10318):16 pp. https://doi.org/201010.1029/2009JA014719

Watson C, White N, Church J, Burgette R, Tregoning P, Coleman R (2011) Absolute Calibration in Bass Strait, Australia: TOPEX, Jason-1 and OSTM/Jason-2. Mar Geodesy 34(3–4):242–260. https://doi.org/10.1080/01490419.2011.584834

Watson C, Legresy B, Beardsley J, King M, Deane A (2018) Absolute altimeter bias results from Bass Strait, Australia. presented at OSTST 2018, Ponta Delgada, Azores

Improvement of the Arctic Ocean Bathymetry and Regional Tide Atlas: First Result on Evaluating Existing Arctic Ocean Bathymetric Models

M. Cancet, O. Andersen, A. Abulaitijiang, D. Cotton, and J. Benveniste

Abstract

The quality of existing bathymetry models for the Arctic Ocean is evaluated through visual comparison and the response of modelled tides. The high resolution ArcTide 2017 hydrodynamic model was used to evaluate the bathymetry in selected shallow water regions where tides are significant. The Southern Barents Sea was identified as a problematic region where inconsistencies were identified, resulting from methods used to patch in regional models and incorrect definitions of coastlines and depths. More generally, the investigation shows that careful verifications are needed to ensure seamless transitions between bathymetry datasets.

More accurate bathymetry in the Arctic Ocean is needed and we outline the development of a new Arctic bathymetry using bathymetry inversion which uses a combination of the existing Arctic bathymetry and topography inverted from a band-pass filtered version of the most recent DTU17 gravity field. We also illustrate the regions where we find adequate spatial correlation to perform such inversion.

Keywords

Arctic Ocean · Bathymetry · Ocean tide · Satellite altimetry

1 Introduction

The CryoSat Plus for Oceans (CP4O) project, under the European Space Agency (ESA) Support to science elements (STSE program, aims to develop and evaluate new ocean products from CryoSat-2 data and so maximize the scientific return of CryoSat-2 over oceans. The main focus of CP4O has been on the additional measurement capabilities that are offered by the Synthetic Aperture Radar (SAR) mode of the Synthetic Aperture Interferometric Radar Altimeter (SIRAL) altimeter, with further work in developing improved geophysical corrections and derived products for the Arctic Ocean.

The Arctic Ocean is a challenging region for such studies because of its complex and poorly documented bathymetry, frequent presence of sea ice, and the scarcity of in-situ observations. One of the objectives of the follow-on of the CP4O project is to produce a new bathymetric dataset for the Arctic Ocean, based on the CryoSat-2 data. The first step of this work consisted of an evaluation of the existing bathymetric datasets in the region in order to choose the basis for building an improved bathymetry from gravity inversion. However, independent observations of bathymetry are scarce in this region. We thus used the fact that tide heights and phases from hydrodynamic tidal models are very sensitive to the bathymetry as the means to evaluate the quality of the different bathymetry datasets, by comparing

M. Cancet
NOVELTIS, Labège, France

O. Andersen (✉) · A. Abulaitijiang
DTU Space, Copenhagen, Denmark
e-mail: oa@space.dtu.dk

D. Cotton
SatOC, Derbyshire, UK

J. Benveniste
ESA/ESRIN, Frascati, Italy

S. P. Mertikas, R. Pail (eds.), *Fiducial Reference Measurements for Altimetry*,
International Association of Geodesy Symposia 150, https://doi.org/10.1007/1345_2019_85

output from high resolution hydrodynamic simulations based on the Arctide2017 regional configuration (Cancet et al. 2018) with tide height coefficients from in situ tide gauges and analyses of satellite altimetry. Arctide2017 is already implemented in the framework of the CP4O initiative. In particular, we identified significant differences between the various bathymetric models in the Southern Barents Sea, with a major problem being an incorrect definition of the coastline in some models.

Better bathymetry is needed and currently there is a substantial global effort called Seabed 2030 (see Mayer et al. 2018) to updated bathymetry both globally and regionally. In Sect. 3 we outline the way a bathymetry model for the deeper part of the Arctic Ocean can be derived using a combination of the Arctic bathymetry we initially find the best, augmented with bathymetry inversion of satellite derived gravity. Satellite derived gravity has developed significantly with the launch and availability of Cryosat-2 SAR altimetry.

2 Arctic Bathymetry Datasets and Errors

We initially assessed the various Arctic Ocean bathymetry data sets, in order to select the best basis for the subsequent improved bathymetry derived using gravity inversion of altimetric gravity.

The three bathymetry datasets investigated here are: the RTopo-1.0.5 bathymetry (Timmermann et al. 2010), the RTopo-2 bathymetry (Schaffer et al. 2016) and the composite bathymetry developed at LEGOS (Laboratoire d'Etudes en Géophysique et Océanographie Spatiales).

All three bathymetric models use the Earth Topography (ETOPO) or General Bathymetric Chart of the Oceans (GEBCO) as the backbone and add in the International Bathymetric Chart of the Arctic Ocean (IBCAO). Subsequently, local modifications based on local surveys are added. Consequently, the models are nearly identical to various versions of the GEBCO/IBCAO bathymetry in large parts of the Arctic Ocean.

The RTopo -1.0.5 bathymetry (Timmermann et al. 2010) is based on the following sources:

> GEBCO (v 2008) at locations poleward of 72° latitude or shallower than 200 m depth, Smith and Sandwell (1997) equatorward of 70° and deeper than 1000 m with smooth blending for areas in-between.

The RTopo-2 bathymetry (Schaffer et al. 2016) is based on the following sources:

> GEBCO 2014 (Weatherall et al. 2015; Becker et al. 2009) and IBCAOv3 (Jakobsson et al. 2012). Fjord and shelf bathymetry around Greenland from Bamber et al. (2013), Arndt et al. (2015) and Seroussi et al. (2011). The specific updates with respect to former IBCAOV3 can be seen in the publication by Schaffer et al. (2016).

The LEGOS composite bathymetry is based on the following data sources (personal communication from F. Lyard):

> ETOPO1 (Amante and Eakins 2009) with IBCAO v2 in the Arctic Ocean as well as Smith and Sandwell (SW-16.1) patches, RTopo-1.0.5 patches, local bathymetric charts for the Laptev, Barents and White Seas.

Note that bathymetry models continue to be updated and a substantial global effort called Seabed 2030 (see Mayer et al. 2018) is ongoing to update GEBCO and IBCAO for the Arctic region.

2.1 Bathymetry Evaluation

The accuracy of bathymetric models can be no better than the accuracy of the data sources from which they are compiled. In very large parts of the Arctic Ocean, the IBCAO model is based on pre-1990 Soviet bathymetry maps. Due to the sparsity of bathymetric data, nearly all available bathymetric surveys have been merged into the IBCAO model. Hence, there is very little independent data to directly assess the different available compiled bathymetries and so it is necessary to consider alternative methods. Two such methods are considered here. One is local visual evaluation and the other is the use of hydrodynamic ocean tide modelling, as a relatively high number of tide gauges are available along the Arctic shorelines and can be used to assess the hydrodynamic simulations.

2.1.1 Bathymetry Evaluation Using Hydrodynamic Tidal Modelling

Hydrodynamic tidal models are very sensitive to the bathymetry accuracy, especially on the shelves where the ocean tides show the largest amplitudes (Padman et al. 2018). In the absence of independent bathymetry data to assess the bathymetry datasets, comparisons between hydrodynamic tidal models and tidal measurements can thus be used as a proxy to evaluate the bathymetry. However, this method is mainly applicable in shallow water regions and preferably in regions where tides are large, like the continental shelves of the Eurasian part of the Arctic Ocean (Kowalik and Proshutinsky 1994; Padman and Erofeeva 2004).

To assess the three Arctic Ocean bathymetry datasets by this method, the TUGO hydrodynamic model was used. This model, developed at LEGOS, has already been used for the implementation of the global models FES2004 (Lyard et al. 2006), FES2012 (Carrère et al. 2012) and FES2014 (Carrère et al. 2015), and for the Arctic regional model Arctide2017 developed in the framework of the CP4O initiative (Cancet et al. 2018). The 2D equations of the TUGO model are based on the classical shallow-water equations of continuity and momentum. The model is run using the finite elements discretization, with triangular elements. The unstructured grid enables the resolution to be increased in the most demanding parts of the model, in terms of coastal geometry and complex hydrodynamic processes. This approach is of particular interest in the shallow waters, where the energy dissipation is the highest due to the bottom friction, as well as in areas of steep topographic gradients, where the currents generally show high variability. and internal tides are generated by the interaction of the barotropic tides with the topography. The conversion of 2D barotropic tides into baroclinic tides are accounted for using a dissipative term.

Three almost identical simulations were carried out, in which the only change was the input bathymetry model. The configuration of the model (e.g., mesh, and bottom friction coefficient) was the same as for the Arctide2017 regional tidal model (Cancet et al. 2018). The three simulations were then compared with tidal harmonic constituents extracted from tide gauges and satellite altimetry observations.

The tide gauge database that we used consists of long-term collation of various historical databases from international programs (International Hydrographic Organization – IHO, World Ocean Circulation Experiment – WOCE) as well as Canadian (Fisheries and Oceans Canada, http://www.isdm-gdsi.gc.ca/isdm-gdsi/twl-mne/index-eng.htm) and Russian (Kowalik and Proshutinsky 1994) sources. The quality of this dataset of about 400 stations is uneven, and the database was strictly edited based on various criteria (dubious locations, clock errors, very short periods, inconsistency with neighbouring stations...). The dataset was finally reduced to 120 stations. For most of them, the elevation time series are not available, which means that the harmonic analysis cannot be run again and cross-checked (see Cancet et al. 2018 for more details on the tide gauge dataset selection). The vector differences, based on the complex tidal amplitudes, between the tidal constituents from the tide gauges and the three simulations are shown for the four largest and dominating tidal components (M2, S2, K1 and O1) in Fig. 1.

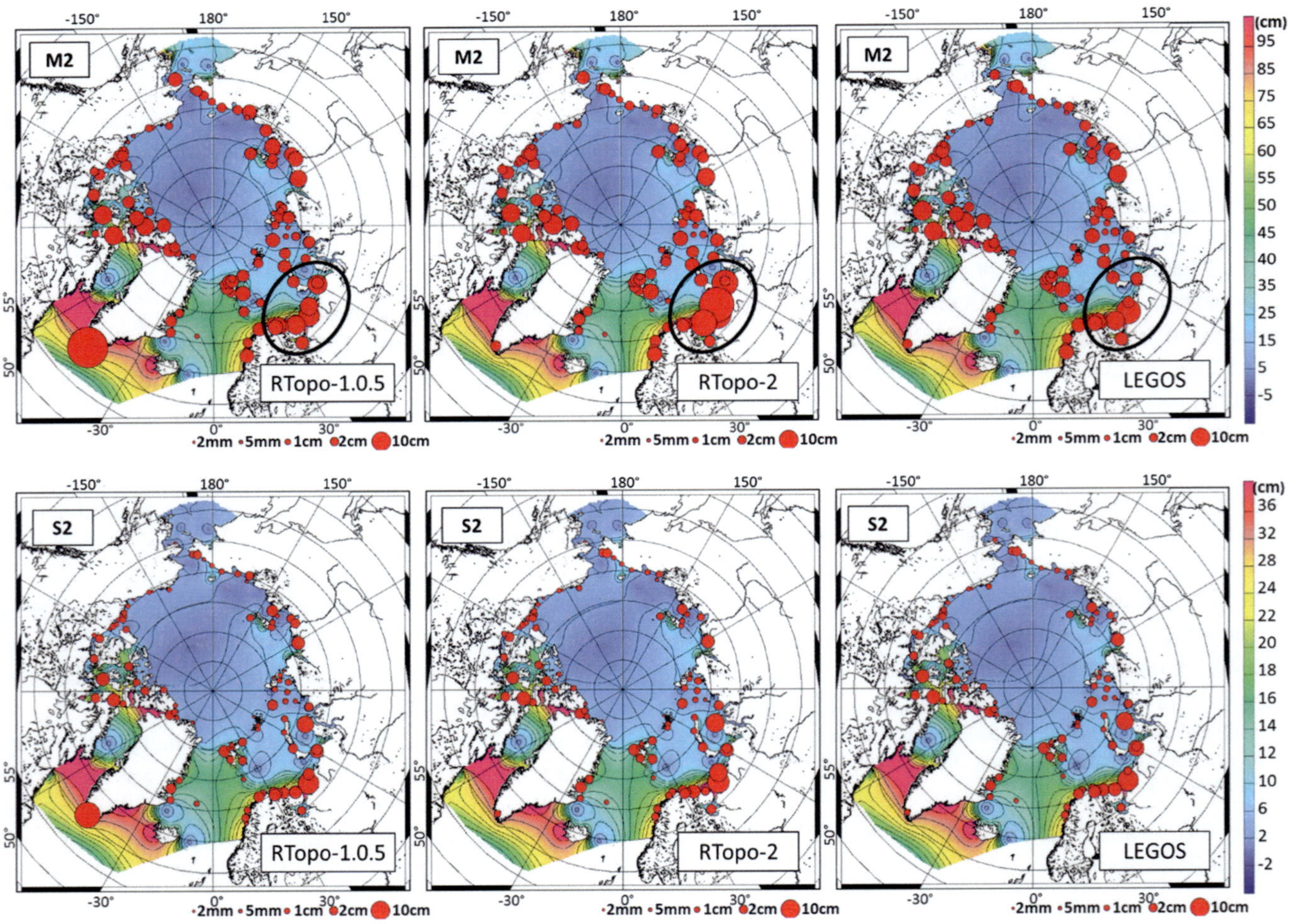

Fig. 1 Misfits (the red dots show the vector differences) between the tide gauges and the three hydrodynamic simulations using different bathymetry RTopo-1.0.5 (right), RTopo-2 (central) and LEGOS (right), for the four main tidal components (M2, S2, K1 and O1). The amplitude of the tidal component is shown in colour in the background for each hydrodynamic simulation. The Barents Sea/Pechora Sea region is encircled in black on the M2 plots

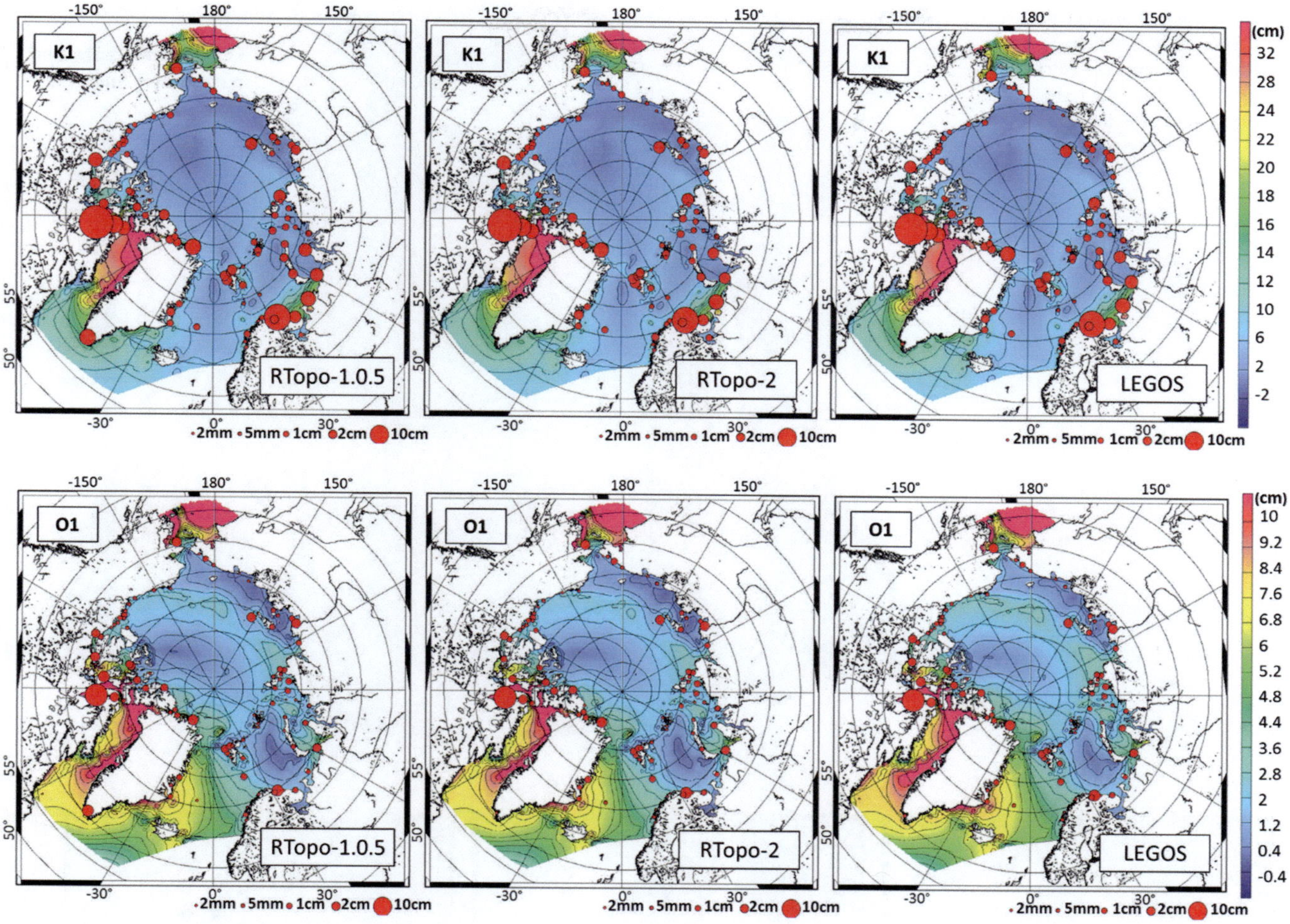

Fig. 1 (continued)

Except for two distinct regions, the performance of all three simulations is nearly identical. A large reduction of the misfit to the tide gauge observations south of Greenland is observed from RTopo-1.0.5 to RTopo-2, for the four tidal components. This can be directly related to the specific improvements made in the Greenland region in the RTopo-2 bathymetry dataset (Schaffer et al. 2016).

Conversely, a large increase in the misfit is seen from RTopo-1.0.5 to Rtopo-2 in the Barents and Pechora Seas for the M2 main tidal component. The bathymetry is shown in Fig. 2 with the Pechora Sea marked as a red rectangle and the entrance to the White Sea as a blue rectangle. Figure 3 shows a zoom in on Fig. 1 in the Pechora Sea, illustrating the misfit with different tide models in this specific region for the M2 tidal component.

Averaged over the Pechora Sea, he model based on LEGOS bathymetry shows the best performance based on comparisons with the tide gauge observations. In the southern part, the increased misfit for RTopo-2 (black ellipse in Fig. 3) is because RTopo-2 is shallower than the other models close to the coast of Russia. This generate errors in the tide modelling. In the narrow strait between mainland Russia and Nova Zemlya (red ellipse in Fig. 3), the RTopo-2 and RTopo-1.5.0 bathymetry datasets exhibit larger misfits to the tide gauges than the LEGOS bathymetry. This is because the strait is closed in RTopo-1.0.5 and extremely shallow in RTopo-2. The strait is considerably wider and deeper in the LEGOS model. The tidal comparison confirms that the strait is wider and deeper similar to the LEGOS model.

Tide gauges are confined to the coast and mainly available in the Russian and Norwegian sectors of the Arctic Ocean. Therefore, we attempted to generalize the approach to support a comparison with tidal constituents determined from altimetry, as presented in Fig. 4. The M2 tidal constituents were determined from 8 years of Envisat and 4 years of Cryosat-2 satellite altimetry missions in boxes of 1° latitude by 3° longitude.

The map of tidal constituent errors (Fig. 4) from satellite altimetry estimates confirms that the LEGOS bathymetry provides the best tide model for the Pechora Sea. This method offers a useful potential for evaluating bathymetric models in regions of the Arctic Ocean where tide gauges are sparse, such as the Canadian Arctic Archipelago.

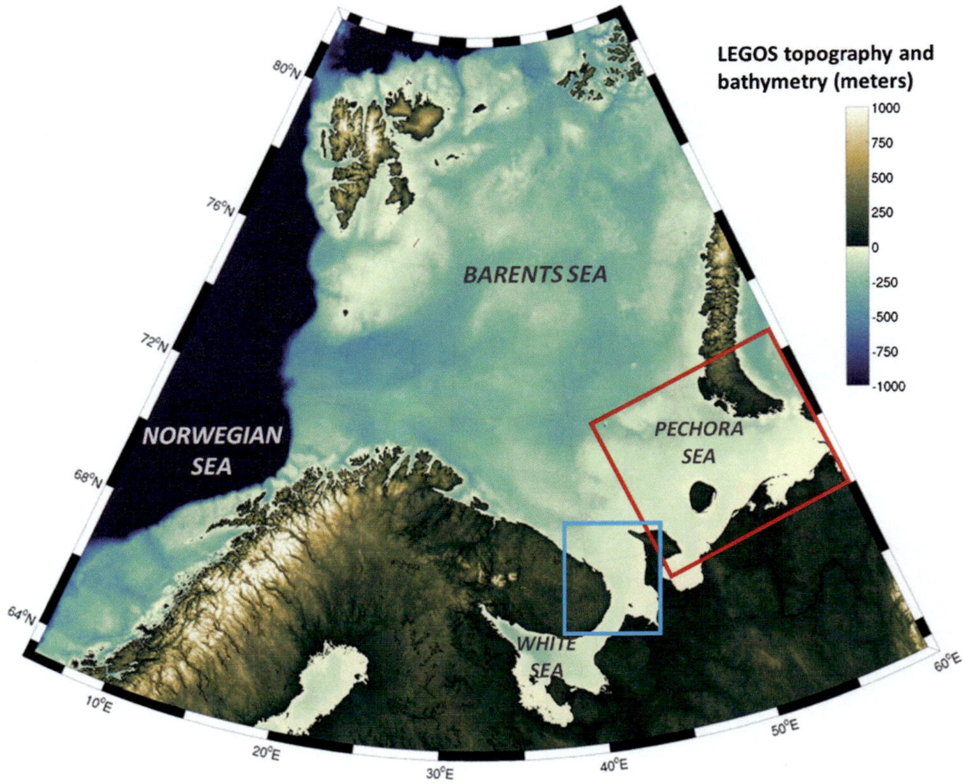

Fig. 2 LEGOS bathymetry of the Barents Sea. The blue rectangle represents the entrance into the White Sea, the red rectangle shows the Pechora Sea

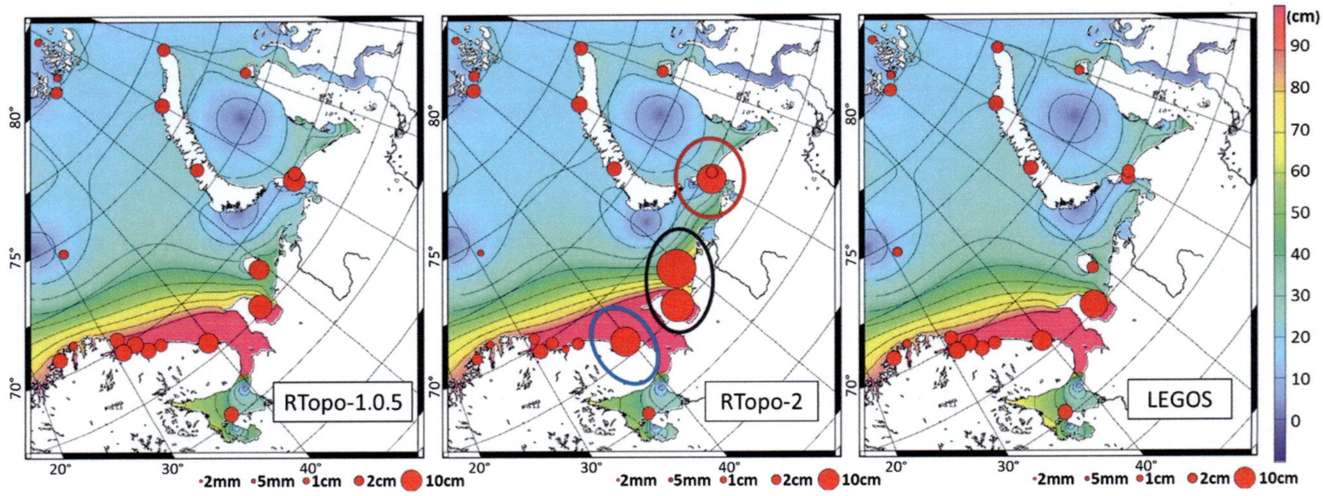

Fig. 3 M2 misfits (vector differences) between the tide gauges and the three hydrodynamic simulations using different bathymetry RTopo-1.0.5 (right), RTopo-2 (central) and LEGOS (right) in the Pechora Sea. The amplitude of the M2 tidal component is shown in colour in the background for each hydrodynamic simulation. The blue, black and red ellipses on the RTopo-2 plot highlight the regions where the simulations differ most

2.1.2 Local Evaluation in the White Sea

The relatively shallow entrance into the White Sea from the Barents Sea (blue rectangle in Fig. 2) illustrates the problem with the way the different Arctic bathymetry models are patched together from various local/global bathymetric models. Figure 3 (blue ellipse) and Fig. 4 show that the misfits between the hydrodynamic simulations and the observations at the entrance of the White Sea are larger using the RTopo-2 bathymetry dataset than for the other models.

The entrance into the White Sea is relatively shallow, less than 60 m. Bathymetry from the three considered bathymetric models is shown in Fig. 5. Several distinct features resulting from recent updates are circled in black. For the RTopo-2 bathymetry, a small, unrealistic cape on the northern coast of the Kola Peninsula, and a 5 m-high "runway" structure stretching 70 km north from the coast of Russia, can be seen. Both features appear in IBCAO V3 and have been propagated into the RTopo-2 bathymetry model (and

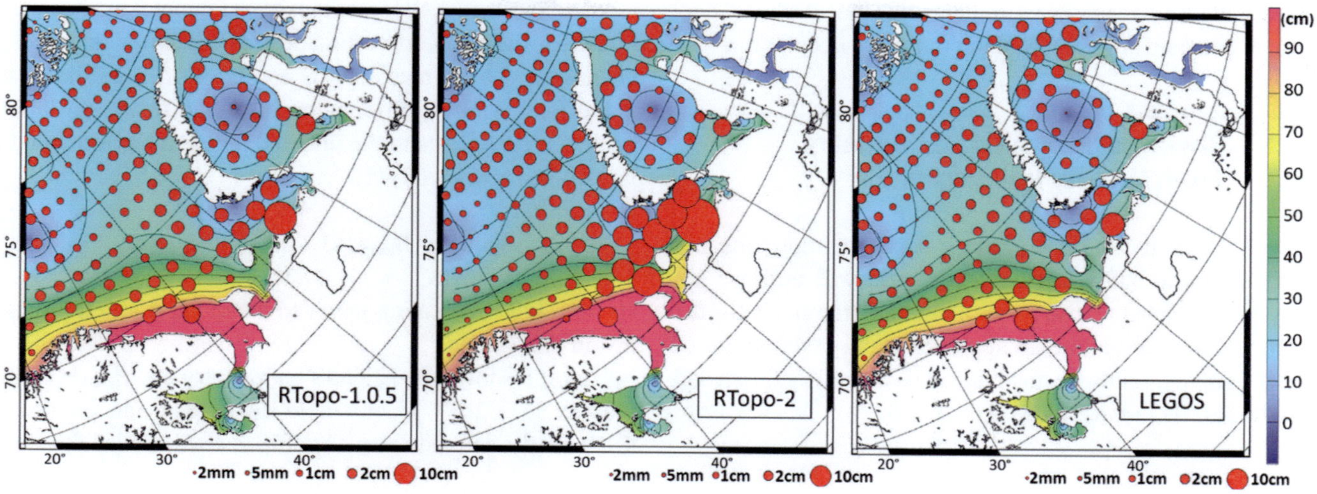

Fig. 4 M2 misfits (vector differences) between altimetry and the three hydrodynamic simulations using different bathymetry RTopo-1.0.5 (right), RTopo-2 (central) and LEGOS (right) in the Pechora Sea. The amplitude of the M2 tidal component is shown in colour in the background for each hydrodynamic simulation

Fig. 5 Bathymetry of the entrance into the White Sea from the Barents Sea from the different bathymetric models. The lower right-hand map shows the Source Identification Grid (SIG) of IBCAOv3. Purple colour indicates data source from land DEM; green implies ship sounding data and orange lines indicate contour lines where no ship soundings exists

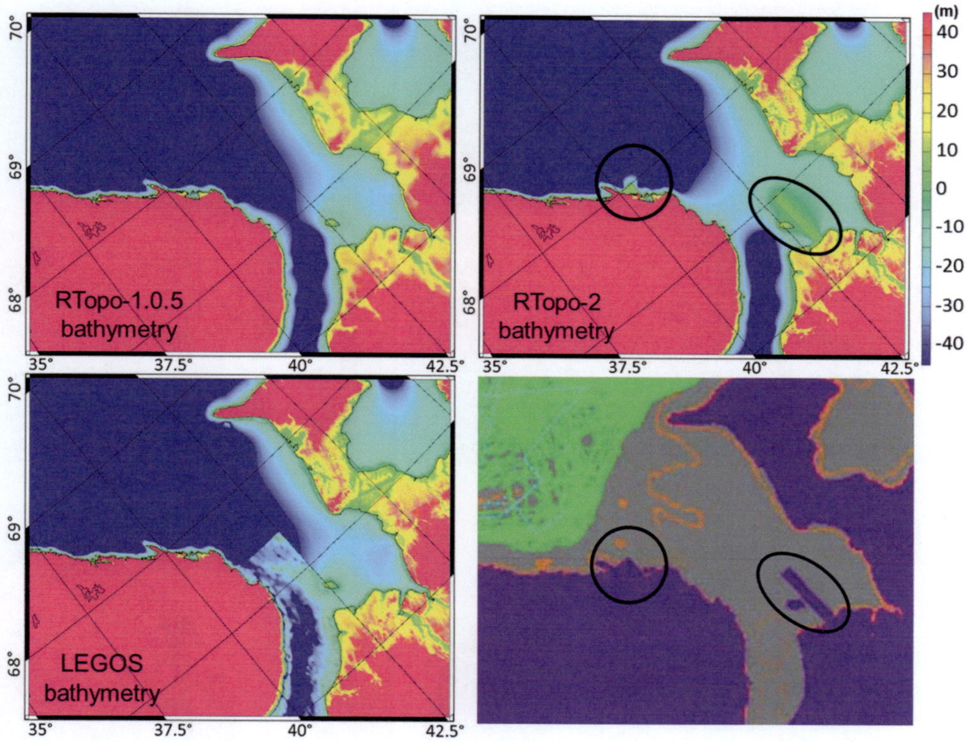

also in the GEBCO 2014 bathymetry model). A closer look at the source information for IBCAO V3 (lower right-hand plot in Fig. 5) identifies both problems as being coding of this region as land (purple data in the figure from land DEM). Both features are also visible in the Smith and Sandwell (SW-16.1) bathymetry which was likely patched into the IBCAO V3 model. The presence of this unrealistic cape in the RTopo-2 bathymetry is directly translated into larger misfits to the nearby tide gauges in the corresponding tidal simulation, as shown in Fig. 3 (blue ellipse).

The LEGOS bathymetry is generated from two different sources of bathymetry in the region: an extract of the Smith and Sandwell bathymetry was patched in the White Sea south of 68°N and west of 42.5°E, while the rest of the bathymetry comes from ETOPO1 in the Barents Sea. This patching results in a step of several meters in the bathymetry. We therefore advocate for very careful verifications to ensure seamless transitions between the various patches in developing bathymetry models.

Further detailed comparisons of land masks and bathymetry maps (not shown here) throughout the Arctic

Ocean identified smaller problems in the Pechora Sea, Laptev Sea, and around the southern tip of Greenland.

3 Towards Improved Bathymetry from Gravity Inversions

The analyses in Sect. 2 point to a need for improving our grids of Arctic bathymetry. A better grid will improve not only tide models but also general circulation models for the Arctic. Here, we outline an approach using assimilation of marine gravity, following the method described by Smith and Sandwell (1997) and Sandwell et al. (2002).

We chose the LEGOS composite bathymetry as the basis model for the altimetric estimation. In the initial approach, the predicted bathymetry $H_p(x)$ can be written as the sum of the long wavelength component of the (existing) input bathymetry $B_{\text{long}}(x)$ and the inverted topography from band-pass filtered gravity $G_{BP}(x)$:

$$\mathbf{H_p\,(x) = B_{long}\,(x) + S\,(x) \bullet G_{BP}\,(x)}$$

where S(x) is the scaling factor to invert gravity to topography, in m/mGal.

Figure 6 shows the LEGOS bathymetry and the recently released DTU2017 gravity field (Andersen and Knudsen 2019) partly derived from CryoSat-2 after spatial filtering

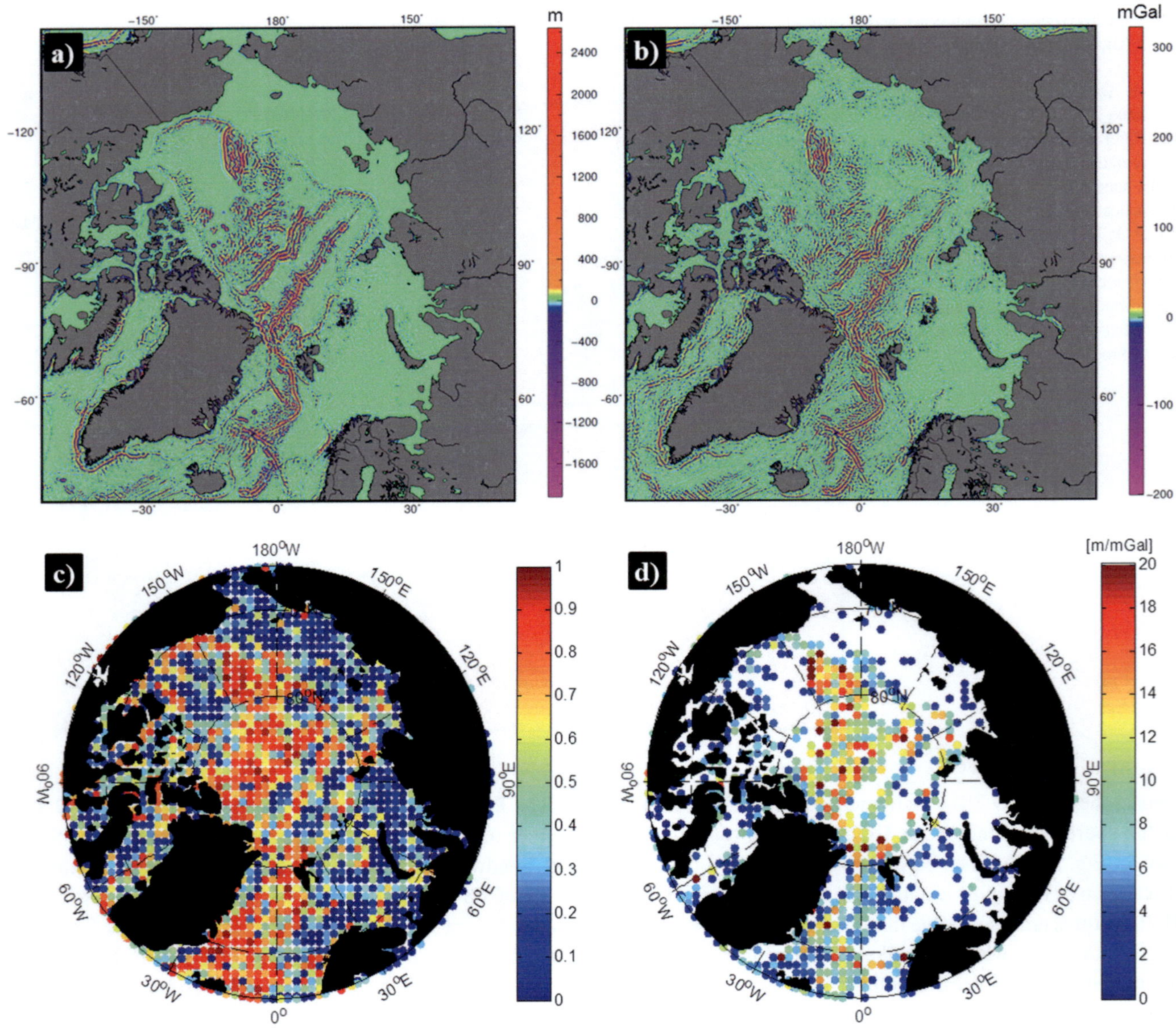

Fig. 6 Band-pass filtered (**a**) LEGOS composite bathymetry and (**b**) DTU17 gravity field derived from CryoSat-2. The band-pass filter passes spatial wavelengths between 20 and 70 km. (**c**) The linear correlation, estimated on 100 km spacing grid. (**d**) Scaling factor S(x). Only grid nodes where linear correlations are larger than 0.5 are shown

using a band-pass filter to extract the signal between 20 km and 70 km spatial wavelengths. We limited the maximum wavelength to 70 km (compared with 160 km in Smith and Sandwell 1997), to suppress the long-wavelength error in gravity models at the high latitudes. We then use the band-pass filtered bathymetry and gravity to estimate the linear correlation and scaling parameter S(x) (gravity-to-topography admittance) on grid nodes spaced 100 km apart.

Moderate coherence and positive scaling factor S(x) between bathymetry and free air gravity for wavelength between 20 km and 70 km is generally found for latitudes above 80° along the Gakkel Ridge, Lomonosov Ridge, Makarov Basin and the North Atlantic Boreas Basin, which is located between Greenland and Svalbard (Björk et al. 2018). However, on most parts of the shallow continental shelf in the Arctic and sedimentary basins (e.g., Canada Basin, Amundsen Basin and Nansen Basin), the correlations are low (Fig. 6c). This is due to the thick sediments (Laske and Masters 1997) on the sea floor, which appears flat on the bathymetry maps. The gravity signal is a combination of sedimentary layer and buried tectonic features beneath the sediments. In such parts, we need to set the scaling factor $S(x) = 0$ to keep the original bathymetry grid values. In Fig. 6d we used a correlation threshold of 0.5 to filter out the low correlation nodes in the S(x) grid.

Further analysis into the bathymetry estimation can be found in Abulaitijiang et al. (2019) who investigated different filtering approaches being more appropriate for the Arctic Ocean compared to the standard filtering applied by Smith and Sandwell (1997). The results by Abulaitijiang et al. (2019) indicate that this approach will lead to the best available Arctic bathymetry for future modelling studies, and help us identify further errors in bathymetry than will indicate priorities for future bathymetry measurements.

4 Conclusions

We have evaluated existing bathymetry models in the Arctic Ocean through two different techniques. High resolution tide modelling, based on the Arctide2017 regional tidal model configuration, was used to evaluate the bathymetric models through their impact on tidal solutions in shallow water regions. Differences between the various bathymetric models in the Southern Barents Sea were identified by the comparison between the hydrodynamic simulations and the tide gauges in the region. Satellite altimetry derived tidal constituents were also considered, leading to consistent conclusions regarding the impact of different bathymetries on tidal solutions in the Pechora Sea. The use of tidal constituents derived from satellite altimetry offers a promising supplement to be applied in remote regions where there are few tide gauges. We also carried out a visual comparison of bathymetry products in the Barents Sea, identifying anomalous features in the bathymetry which were related to the way that different data sources are patched together to create the different models.

The evaluation of currently available bathymetry models is an essential first step to support the derivation of a new altimetric based Arctic bathymetry, which is based on a band-pass inversion of the most recent high resolution DTU17 gravity field. Here we have outlined the approach that we will take, presenting the first results to identify the regions where the correlation between band-passed bathymetry and altimetric gravity is sufficiently high to support the use of bathymetric inversion. These regions are typically in the deeper parts of the Arctic Ocean. We propose that our combination of bathymetry improvements through gravity inversion, tide height mapping with radar altimeter satellites, and hydrodynamic tide modelling will lead to significant improvements in our ability to model the Arctic Ocean bathymetry and it rapidly changing ice cover. Bathymetry improvement through gravity inversion clearly shows the best result in the deep ocean where we find adequate spatial correlation between bathymetry and gravity. It is therefore interesting if the green laser on ICESAT-2 (icesat-2.gsfc.nasa.gov) can measure bathymetry in the shallow water regions.

Data Availability The DTU17 gravity field and bathymetry models are available from ftp.space.dtu.dk/pub/DTU17 or through request to the authors (oa@space.dtu.dk).

Acknowledgement The authors would like to acknowledge the ESA contribution to the research, through the Cryosat Plus for Oceans program supported under the ESA Support to Science Element Programme (contract 4000106169/12/I-NB).

References

Abulaitijiang A, Andersen OB, Sandwell D (2019) Improved arctic ocean bathymetry derived from DTU17 gravity model. Earth Space Sci 6. https://doi.org/10.1029/2018EA000502

Amante C, Eakins BW (2009) ETOPO1 1 Arc-minute global relief model: procedures, data sources and analysis. In: NOAA Technical Memorandum NESDIS NGDC-24. National Geophysical Data Center, NOAA, Boulder. https://doi.org/10.7289/V5C8276M

Andersen OB, Knudsen P (2019) The DTU17 global marine gravity field – first validation results. In: International association of geodesy symposia. Springer Nature, Basel. https://doi.org/10.1007/1345_2019_65

Arndt JE, Jokat W, Dorschel B, Myklebust R, Dowdeswell JA, Evans J (2015) A new bathymetry of the Northeast Greenland continental shelf: constraints on glacial and other processes. Geochem Geophys Geosyst 16:3733–3753. https://doi.org/10.1002/2015GC005931

Bamber JL, Griggs JA, Hurkmans RTWL, Dowdeswell JA, Gogineni SP, Howat I, Mouginot J, Paden J, Palmer S, Rignot E, Steinhage D (2013) A new bed elevation dataset for Greenland. Cryosphere 7:499–510. https://doi.org/10.5194/tc-7-499-2013

Becker JJ, Sandwell DT, Smith WHF, Braud J, Binder B, Depner J, Fabre D, Factor J, Ingalls S, Kim S-H, Ladner R, Marks K, Nelson S, Pharaoh A, Trimmer R, Rosenberg JV, Wallace G, Weatherall P (2009) Global bathymetry and elevation data at 30 arc seconds resolution: SRTM30_PLUS. Mar Geod 32:355–372. https://doi.org/10.1080/01490410903297766

Björk G, Jakobsson M, Assmann K, Andersson LG, Nilsson J, Stranne C, Mayer L (2018) Bathymetry and oceanic flow structure at two deep passages crossing the Lomonosov Ridge. Ocean Sci 14:1–13

Cancet M, Andersen O, Lyard F, Cotton D, Benveniste J (2018) Arctide2017, a high-resolution regional tidal model in the Arctic Ocean. Adv Space Res 62(6):1324. https://doi.org/10.1016/j.asr.2018.01.007

Carrère L, Lyard F, Cancet M, Guillot A, Roblou L (2012) FES2012: a new global tidal model taking advantage of nearly twenty years of altimetry. In: Proceeding of the 20 years of progress in radar altimetry symposium, Venice, Italy

Carrère L, Lyard F, Cancet M, Guillot A, Picot N, Dupuy S (2015) FES2014: A new global tidal model. Presented at the Ocean Surface Topography Science Team meeting, Reston, USA

Jakobsson M, Mayer L, Coakley B, Dowdeswell JA, Forbes S, Fridman B, Hodnesdal H, Noormets R, Pedersen R, Rebesco M, Schenke HW, Yarayskaya Z, Accettella D, Armstrong A, Anderson RM, Bienhoff P, Camerlenghi A, Church I, Edwards M, Gardner JV, Hall JK, Hell B, Hestvik O, Kristoffersen Y, Marcussen C, Mohammed R, Mosher D, Nghiem SV, Pedrosa MT, Travaglini PG, Weatherall P (2012) The International Bathymetric Chart of the Arctic Ocean (IBCAO) version 3.0. Geophys Res Lett 39:L12609. https://doi.org/10.1029/2012GL052219

Kowalik Z, Proshutinsky AY (1994) The Arctic Ocean Tides. In: Johannessen OM, Muench RD, Overland JE (eds) The Polar Oceans and their role in shaping the global environment. American Geophysical Union, Washington, D.C. https://doi.org/10.1029/GM085p0137

Laske G, Masters G (1997) A global digital map of sediment thickness. EOS Trans AGU 78:F483

Lyard F, Lefèvre F et al (2006) Modelling the global ocean tides: a modern insight from FES2004. Ocean Dyn 56:394–415

Mayer L, Jakobsson M, Allen G, Dorschel B, Falconer R, Ferrini V, Lamarche G, Snaith H, Weatherall P (2018) The Nippon Foundation—GEBCO Seabed 2030 Project: the quest to see the World's Oceans completely mapped by 2030. Geosciences 8:63. https://www.mdpi.com/2076-3263/8/2/63

Padman L, Erofeeva S (2004) A barotropic inverse tidal model for the Arctic Ocean. Geophys Res Lett 31:L02303. https://doi.org/10.1029/2003GL019003

Padman L, Siegfried MR, Fricker HA (2018) Ocean tide influences on the Antarctic and Greenland ice sheets. Rev Geophys 56:142–184. https://doi.org/10.1002/2016RG000546

Sandwell DT, Gille ST, Smith WHF (2002) Bathymetry from space: oceanography, geophysics, and climate, Bethesda, MD, USA, pp 1–24

Schaffer J, Timmermann R, Arndt J, Kristensen SS, Mayer L, Morlighem M, Steinhage D (2016) A global, high-resolution data set of ice sheet topography, cavity geometry, and ocean bathymetry. Earth Syst Sci Data 8(2):543–557. https://doi.org/10.5194/essd-8-543-2016

Seroussi H, Morlighem M, Rignot E, Larour E, Aubry D, Dhia HB, Kristensen SS (2011) Ice flux divergence anomalies on 79north Glacier, Greenland. Geophys Res Lett 38:L09501. https://doi.org/10.1029/2011GL047338

Smith WHF, Sandwell DT (1997) Global sea floor topography from satellite altimetry and ship depth soundings. Science 277(5334):1956–1962. https://doi.org/10.1126/science.277.5334.1956

Timmermann R, Le Brocq A, Deen T, Domack E, Dutrieux P, Galton-Fenzi B, Hellmer H, Humbert A, Jansen D, Jenkins A, Lambrecht A, Makinson K, Niederjasper F, Nitsche F, Nøst OA, Smedsrud LH, Smith WHF (2010) A consistent data set of Antarctic ice sheet topography, cavity geometry, and global bathymetry. Earth Syst Sci Data 2:261–273. https://doi.org/10.5194/essd-2-261-2010

Weatherall PK, Marks K, Jakobsson M, Schmitt T, Tani S, Arndt JE, Rovere M, Chayes D, Ferrini V, Wigley R (2015) A new digital bathymetric model of the world's oceans. Earth Space Sci 2:331–345. https://doi.org/10.1002/2015EA000107

Sea Level Trends and Variability in the Adriatic Sea and Around Venice

Stefano Vignudelli, Francesco De Biasio, Andrea Scozzari, Stefano Zecchetto, and Alvise Papa

Abstract

A preliminary analysis of sea level (SL) changes around Venice from three tide gauges (TGs) (one off-shore: AAPTF, one at the coast: DSL, and one inside the lagoon: PS) to characterize the variability during 1993–2015 and relative SL trends, is provided. As no global positioning system (GPS) data covering the same period was available to the authors for the three tide gauges, the analysis is restricted to changes relative to the land. Monthly SL means from the European Space Agency (ESA) Sea Level Climate Change Initiative (CCI) altimeter-derived product are also used. A comparison between the monthly mean time series of CCI and AAPTF has been performed using the nearest CCI grid point to the location of AAPTF: the centered Root Mean Square Difference (cRMSD: the root mean square difference of the two series with the respective means removed) resulted 6.33 cm, while the Pearson's linear correlation reached 0.75. Much higher agreement was found, as expected, between the monthly mean records of AAPTF and PS TGs: the cRMSD was 1.03 cm, and the linear correlation 0.99. We obtained a trend of 6.65 mm year^{-1} at AAPTF over the Satellite Altimetry (SA) era (1993–2015). A smaller trend has been found here from altimetry (4.25 mm year^{-1}). The differences might be explained in terms of Vertical Land Motion (VLM) which was not accounted for in the TGs time series, to the different processing of TGs and altimeter data (in the altimeter signal the Dynamic Atmospheric Correction is removed), and/or uncertainties in this area due to the current CCI product that is based on open ocean altimetry. In general, the altimetry trends derived from CCI are spatially higher in the Adriatic Sea than Global Mean Sea Level (GMSL) in most of the region, with greater values than average in Venice. Reprocessing of along-track altimeter data sets with consistent coastal processing for all missions is expected to enhance SL accuracy and with a better refining of raw trends. The SA era is too short to delineate or discuss an affordable climatology, as decadal SL variations cannot be accounted for by such a short time series, nonetheless the rates calculated in this study are fruitfully compared each other and with those derived by SA.

Keywords

Adriatic Sea · Coastal altimetry · Satellite radar altimetry · Sea level · Tide gauge · Venice

S. Vignudelli (✉)
Consiglio Nazionale delle Ricerche (CNR-IBF), Pisa, Italy
e-mail: vignudelli@pi.ibf.cnr.it

F. De Biasio
Consiglio Nazionale delle Ricerche (CNR-ISMAR), Venezia, Italy

A. Scozzari
Consiglio Nazionale delle Ricerche (CNR-ISTI), Pisa, Italy

S. Zecchetto
Consiglio Nazionale delle Ricerche (CNR-ISAC), Padova, Italy

A. Papa
Centro Previsioni e Segnalazioni Maree – CPSM, Venice, Italy

S. P. Mertikas, R. Pail (eds.), *Fiducial Reference Measurements for Altimetry*,
International Association of Geodesy Symposia 150, https://doi.org/10.1007/1345_2018_51

1 Introduction and Study Area

The Adriatic Sea, including the area around Venice, is an important laboratory for validating and applying satellite altimetry products since the calibration of the European Space Agency (ESA) ERS-1 altimeter (Scharroo 2002). The area is frequently affected by storm surges because of its geographical position and geometry. The Sirocco wind blowing along the main axis of the basin rises up water in the Northern Adriatic that generates surges along the northern shorelines. The resulting surge signal is maximum in the northwestern part of the basin, especially at the city of Venice and its surrounding lagoon, where is it called "Acqua Alta" (high water). Also the northeastern side of the Adriatic Sea is subject to storm surges, mainly connected with the phenomenon of meteotsunami, a tsunami-like event enhanced and conditioned by specific meteorological forcing and geometry of the coast (Orlić 2015).

The City of Venice is today on average 30 cm lower than early 1900s (Cordella et al. 2010; Collini et al. 2017). In the Fifth Assessment Report (AR5) of the Intergovernmental Panel on Climate Change, Church et al. (2013) predicted an increase in sea level (SL) between 18 and 59 cm during the next 100 years. The Sixth Assessment Report (AR6) is underway and a special report on the Ocean and Cryosphere in a Changing Climate will look at SL rise and implications for low lying islands, coasts and communities.

In Venice, the MOdulo Sperimentale Elettromeccanico (MOSE) barriers, due for completion in late 2021, were designed to protect the town and its lagoon from storm surges, and preserve its integrity from the ever more frequent high waters connected to the local eustatism and subsidence. The barriers will be raised, closing the mouths only when levels of 110 cm referred to the Zero Mareografico di Punta Salute (ZMPS) are foreseen. ZMPS is the historical reference benchmark of the Mean Sea Level (MSL) in Venice for 25 years 1885–1909. That value is the maximum level that still permits almost normal activities in town. ZMPS is about 30 cm lower than the local mean sea level of the last 15 years. MOSE has been designed to work with an increase of up to 60 cm. Flooding could be further aggravated by the potential rise in SL resulting from climate change.

Understanding SL variability, in particular the relationship between storm surges and long-term SL changes as well as a better assessment of the climate-related contribution to local SL has particular relevance for local policymakers. It is thus of particular interest to have accurate SL observations to improve the scientific knowledge of past trends that might continue into the future. Climate-related changes of meteorological and marine variables are an important input to define future scenarios not only of MOSE infrastructure, designed to protect Venice from storm surges, but also of the coastal settlements of the northern Adriatic Sea. In this context an accurate definition of the actual SL rise cannot leave aside the estimation of the Vertical Land Motion (VLM), as even little rates of it can exacerbate the effects of the SL rise and quicken the attainment of specific thresholds requiring actions in the coastal zone local planning.

Satellite radar altimetry is designed to measure height profiles at sea in the along-track direction, with a coverage organized as a mesh of ground tracks, including also zones with no tide gauges (TGs) around. Until a few years ago, satellite radar altimetry was not sufficiently exploited for SL research in the Adriatic Sea. The ESA eSurge-Venice project permitted to exploit altimeter data to improve the modelling and forecasting of storm surges (De Biasio et al. 2016, 2017; Bajo et al. 2017). The ESA Climate Change Initiative (CCI) now provides a SL Essential Climate Variable from which to get a much clearer picture of regional trends and year-to-year variations (Legeais et al. 2018).

This work reports on the initial assessment of the quality of the ESA SL CCI products in the Adriatic Sea and in particular around the city of Venice (Fig. 1). TGs available around Venice and Trieste provide an accurate independent source of SL information to be used as reference of long-term SL variability at the coast. In this work, we focus on SL changes at long term scales (trends).

2 Data and Methods

A TG records how the SL is changing relative to the land on which it is installed. SL is generally provided as a measurement relative to an arbitrary and local reference that is, by convention, the ZMPS defined previously. The Venice tide gauge installed at "Punta della Salute", indicated as "VENICE PS" or simply "PS" henceforth, supplies one of the longest SL records of the Mediterranean Sea. The SL time series of the historical TG of VENICE PS dates back to about 1875.

The VENICE PS TG is operated by the Venice Tide Centre (Centro Previsioni e Segnalazioni Maree – CPSM, Venice Municipality). As it is situated in the city centre inside the Venice Lagoon, far from any satellite altimeter track, we adopted for this study the SL record supplied by the CPSM TG installed in the ACQUA ALTA off-shore platform (AAPTF: 45° 18′ 51.29″ N, 12° 30′ 29.69″ E) of Consiglio Nazionale delle Ricerche (CNR, National Research Council) (see Fig. 1). The AAPTF record begins in 1974 and is still continuing. It is very useful for the forecast of storm surges in Venice as the SL signal is observed about 50 min in advance with respect to VENICE PS. As the AAPTF record presents gaps in the satellite altimetry era (e.g., 1993 is almost completely missing), we used the SL record registered in another TG of the CPSM, namely VENICE Diga Sud Lido,

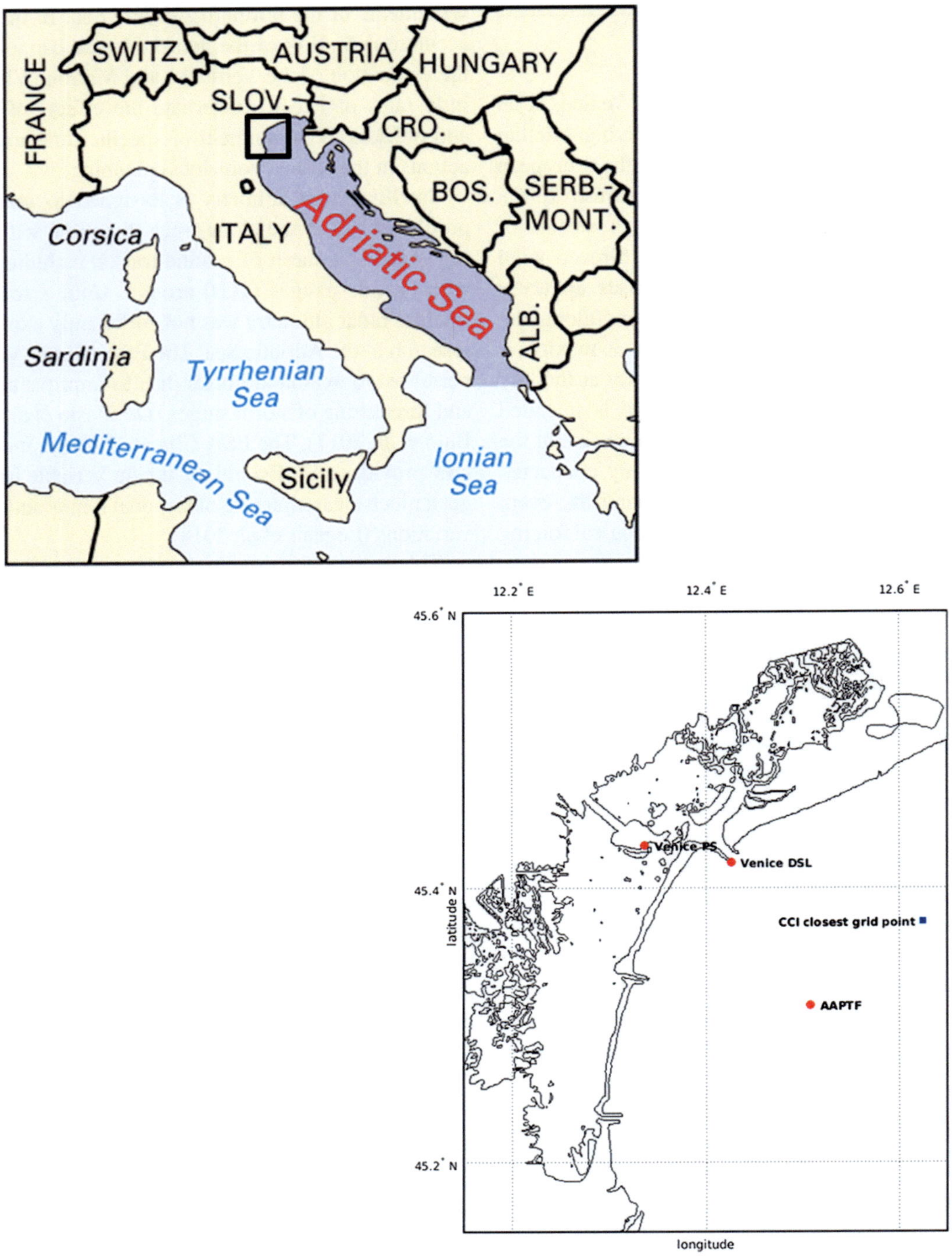

Fig. 1 The Adriatic Sea and in particular the area of investigation (northern part) around City of Venice. Selected tide gauges: CNR Acqua Alta Platform (AAPTF), Venezia Punta della Salute (VENICE PS) and Venezia Diga Sud Lido (VENICE DSL). Also shown the closest CCI grid point to AAPTF

indicated as "DSL", whose position is also shown in the map of Fig. 1, to integrate the AAPTF record. The three TGs cited so far are well maintained and constantly monitored, and report the "equivalent" SL height observed at VENICE PS. Moreover, the AAPTF and DSL TGs have very similar average delays of the tide (Ferla et al. 2007) with respect to PS, thus facilitating the integration of the two time series where needed. The assumption is supported by the plots in Fig. 2. The hourly values of the two TGs have been processed in monthly aggregation to determine the centred (unbiased) Root Mean Square Difference (cRMSD), the Pearson's linear correlation coefficient R and the bias of the two SL time series, for every month, from 1974 to 2016. The results of each month of the 12 months of a whole year are plotted,

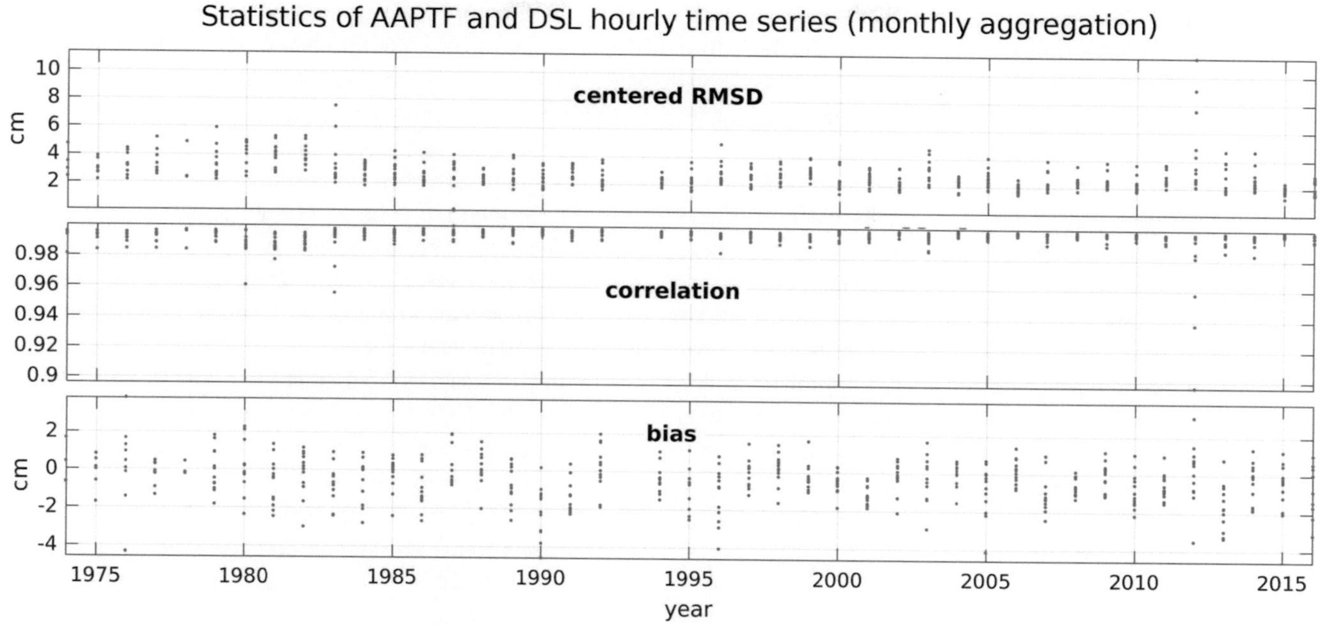

Fig. 2 Centred root mean square difference, linear correlation and bias of the AAPTF and VENICE DSL tide gauge records. Monthly statistics of hourly data

vertically, at the correspondent year, so that each year's seasonal cycle in the three parameters is visually represented along a vertical segment at the given year. The seasonal variability of the three parameters is rather limited across the 43 years: cRMSD ranges from 2 to 5 cm; the linear correlation is contained in the range [0.98, 0.99]; the bias lies in the interval [−2, +2] cm. Only 1980, 1983 and 2010 show a higher variability for some of the three parameters. In 1993 there was not enough data for the analysis. The longest gap in the AAPTF record is the whole 1993 year. Looking at the three plot in the years immediately before (1991, 1992) and after (1994, 1995) the year 1993, the three statistical parameters are well inside the ranges given above. Therefore, the replacement of the entire year of sea level measurements at AAPTF with those taken at DSL seemed to us a reasonable choice. Shorter gaps were filled in the same way, making similar assumptions. In order to be comparable to satellite altimetry records, the TGs hourly records have been reduced to daily means applying the Doodson's X0 filter (Shirahata et al. 2016) following Permanent Service Mean Sea Level guidelines. The daily means were then used to calculate the monthly means.

Since 1990s, a series of radar altimetry missions accumulated a satellite-based record of SL that is now long enough to estimate trends. Sea Level Anomaly (SLA) is obtained by subtracting the Mean Sea Surface (MSS) to the Sea Surface height (SSH), referred to ellipsoid, corrected for various effects (including tides and meteorological forcing to avoid aliasing). The ESA CCI project on "Sea Level" has reprocessed these altimeter data over 1993–2015 to provide homogeneous SL for all altimetry missions (Legeais et al. 2018). The v2.0 dataset was released in December 2016, with details provided at http://www.esa-sealevel-cci.org/products (Quartly et al. 2017). The ESA SL CCI products are generated using open ocean altimetry data, selecting improved satellite orbits and updated geophysical corrections, adopting a new calculation of the mean sea surface used as reference, reducing instrumental drifts and biases, in order to further reduce the error budget and provide a consistent unbiased SL record for long-term change studies. These products include along-track SLA at 1 Hz (around 7 km) and monthly gridded time series of multi-mission merged SLA at a spatial resolution of 0.25° (around 25 km) from which some oceanic indicators (e.g., trends) are derived. The gridding process is described in Ablain et al. (2017).

As the CCI dataset is produced with the Dynamic Atmospheric Correction (DAC) applied, we decided to form another dataset of TGs observations with the Inverse Barometer (IB) (Wunsch and Stammer 1997) effect subtracted. The IB correction is formed by subtracting from the observed sea level at the tide gauge the contribution due to the atmospheric loading due to the difference between the local pressure and the global mean over the oceans. This contribution (in cm) is proportional to a constant (−0.9916 cm mbar^{-1}) by the pressure difference in millibar:

$$\eta^{ib} = -0.9916\left(P_a - \overline{P_a}\right)$$

An hourly time series of atmospheric pressure in Venice has been supplied by Venice Tide Centre. The time series of monthly average atmospheric pressure over the oceans were obtained from the ERA Interim reanalysis project (Berrisford et al. 2011). The period of interest (1993–2015) is well covered by ERA Interim, and the 80 km resolution is adequate to perform a world average. From the time series of the local and the world oceans monthly mean pressure differences, the IB contribution near Venice was finally calculated, and the IB corrected trends are also reported in the text, where needed. Wherever trends are involved in the text, their errors are calculated as standard trend errors, and autocorrelations has not been taken into account. For this reason the error estimates on trends have to be considered optimistic at this stage. As a first rough evaluation, autoregressive coefficients in the monthly SL time series yield values of about 0.4, thus a more realistic estimate of the trends errors could be obtained multiplying the error calculated with linear regression by 1.5 (Von Storch and Zwiers 2001).

3 Results

As an example, the raw time series of SL hourly values at VENICE PS is shown in Fig. 3 (upper panel). The blue line is SL. The astronomical tide contribution (red line) and the residual difference (green line) after subtracting it from the SL are also reported. In order to compare the time series of in situ data to the satellite-based SL measurements, the exact composition of the atmospheric corrections applied to the altimetry record has to be carefully considered, as the altimeter path delay explained by other climatological variables might have been already accounted for, as the MSL atmospheric pressure (Fig. 3, bottom panel) and the wind stress. The atmospheric loading is well observed by TG measurements: 1 mbar increase of the local MSL pressure roughly corresponds to a drop of 1 cm in the SL, owing to the inverse barometer effect.

Monthly means of SL height at the AAPTF and VENICE PS during the CCI reprocessed period (1993–2015) are shown in Fig. 4. The trends are also indicated as dashed lines. The references are arbitrary for sake of readability. VENICE PS and AAPTF overlap pretty well: cRMSD is 1.03 cm, and the Pearson's linear correlation coefficient R reaches 0.99. Recent trends (6.29 ± 1.53 mm year^{-1} at VENICE PS and 6.65 ± 1.50 mm year^{-1} at AAPTF) are more marked with respect to those calculated from the whole datasets (1941–2017 and 1975–2016 respectively) and not showed here. TG measurements register VLMs that are mixed with SL variations. After removal of the IB contribution, calculated as described in the "Data and methods" section, the cRMSD and the Person's linear correlation coefficient between AAPTF

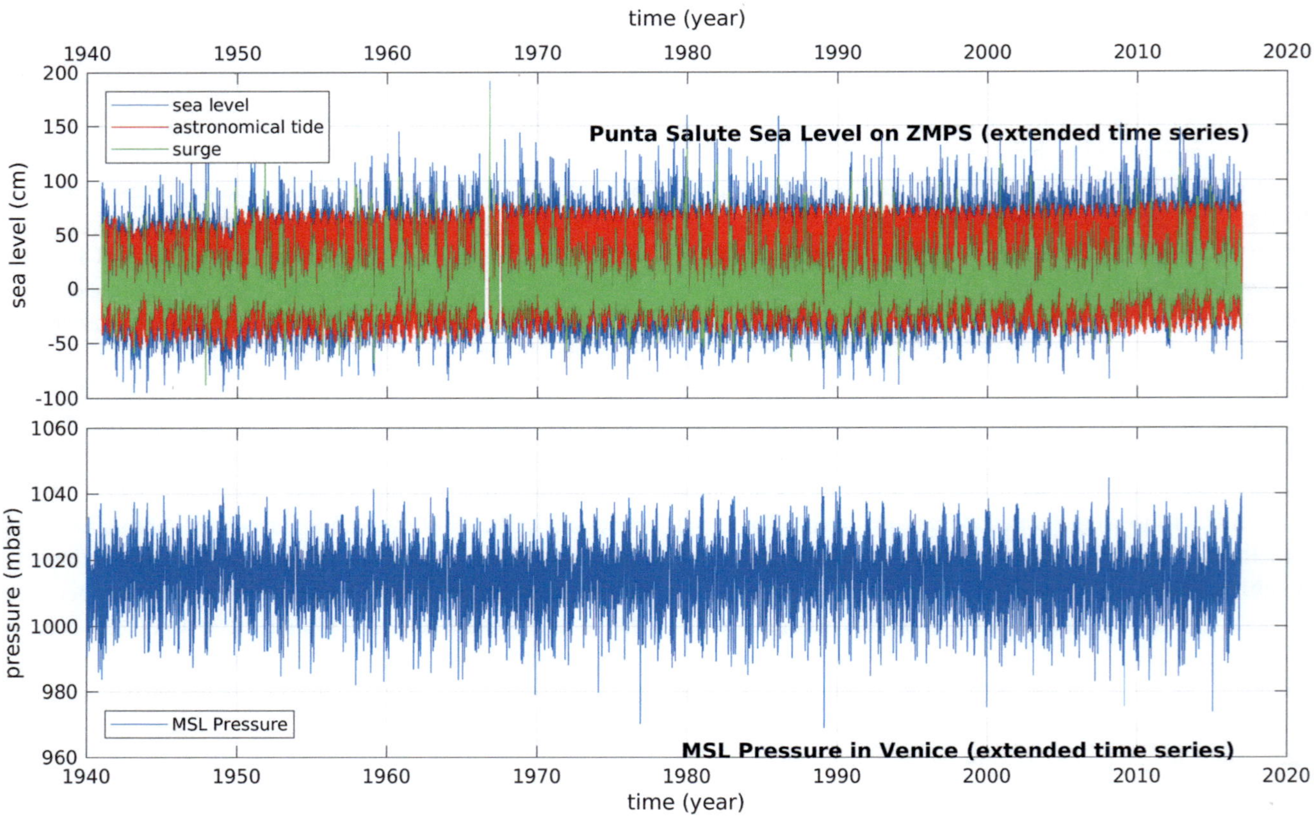

Fig. 3 VENICE PS record. Top: tide gauge hourly time series. Blue: sea level. Red: astronomical tide. Green: difference. Bottom: mean sea level pressure

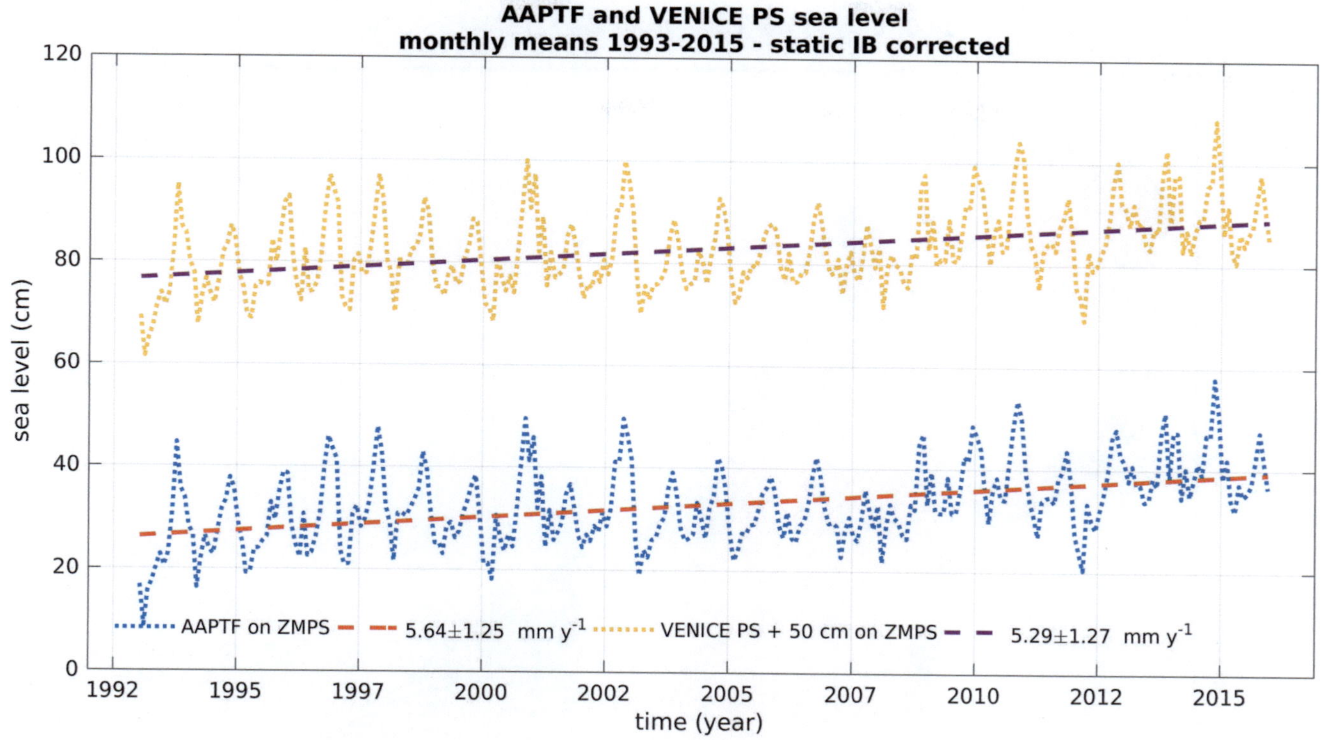

Fig. 4 Sea level monthly means records of AAPTF and VENICE PS during the altimetry era (1993–2015). References are arbitrary for sake of readability

and PS remain the same, while the trends result lower: AAPTF 5.65 ± 1.25 cm, and PS 5.29 ± 1.27 cm. The lower standard deviations on the IB corrected slopes reflect the fact that the IB contribution explains part of the variability of the TG signals, and its removal from the SL monthly time series is thus advisable in order to compare the TGs data with those derived from the CCI altimetry dataset, where DAC, which includes the IB effect, has been already removed.

From the perspective of satellite altimetry, sea level observations highlight that rising is not geographically uniform worldwide (Nerem et al. 2018). A marked spatial variability of SL trends is observed in the Mediterranean Sea (Bonaduce et al. 2016). Some negative trends observed in the Ionian Sea and south-east of Crete, are supposed related to important changes in the circulation observed since 1990s. In the Adriatic Sea, trends are spatially higher than Global Mean Sea Level (GMSL) in most of the region, with a pronounced greater value than average around Venice, as shown in Fig. 5.

A comparison has been performed using the nearest CCI grid point to the location of AAPTF (Fig. 2). As the CCI dataset is available during the time frame 1993–2015, monthly time series and trends of SLA (altimetry) and SL (tide gauge) are referred to this period (Fig. 6). Both data sets overlap pretty well: cRMSD between AAPTF SL and the altimetry SLA is 6.35 cm for PS, and 6.33 cm for AAPTF, The linear correlation coefficients are also similar: R is 0.74 for AAPTF and 0.75 for PS. Far better results are obtained after removal of the IB effect from the TGs' SL time series: AAPTF-CCI and PS-CCI reach a centred RMSD of 4.00 cm and 3.99 cm respectively (a net 37% reduction), while the linear correlations increase up to 0.87 (+16%) for both.

In term of seasonality, after removal of the IB effect from the TGs time series, the mean seasonal cycles of the TG SL and the altimetry SLA peak almost at the same time (minima in March–April and maxima in November). The three seasonal cycles present almost the same amplitude (peak to peak): 16 cm for SLs, and 15 cm for SLA. The three seasonal cycle never differ by more than 2 cm (Fig. 7).

As shown previously, 5.65 mm year^{-1} are measured at AAPTF. A smaller trend has been found from altimetry (4.25 mm year^{-1}). Both trends are much higher than GMSL which is 3.3 mm year^{-1} (Legeais et al. 2018). Fenoglio-Marc et al. (2012) found around Venice 5.6 ± 1.6 mm year^{-1} over a shorter period (1993–2008) and using only Envisat, TOPEX/Poseidon and Jason-1/2 satellite missions. Rocco (2015) found 4.02 ± 1.18 mm year^{-1} from a TG inside lagoon: however, the SL measured by TG was corrected for VLM over the CCI period 1993–2013. It is important that the GPS time frame overlaps completely with the CCI data record.

The difference between TG and altimetry time series is very low. In terms of slopes, the fitting lines differ by a mere

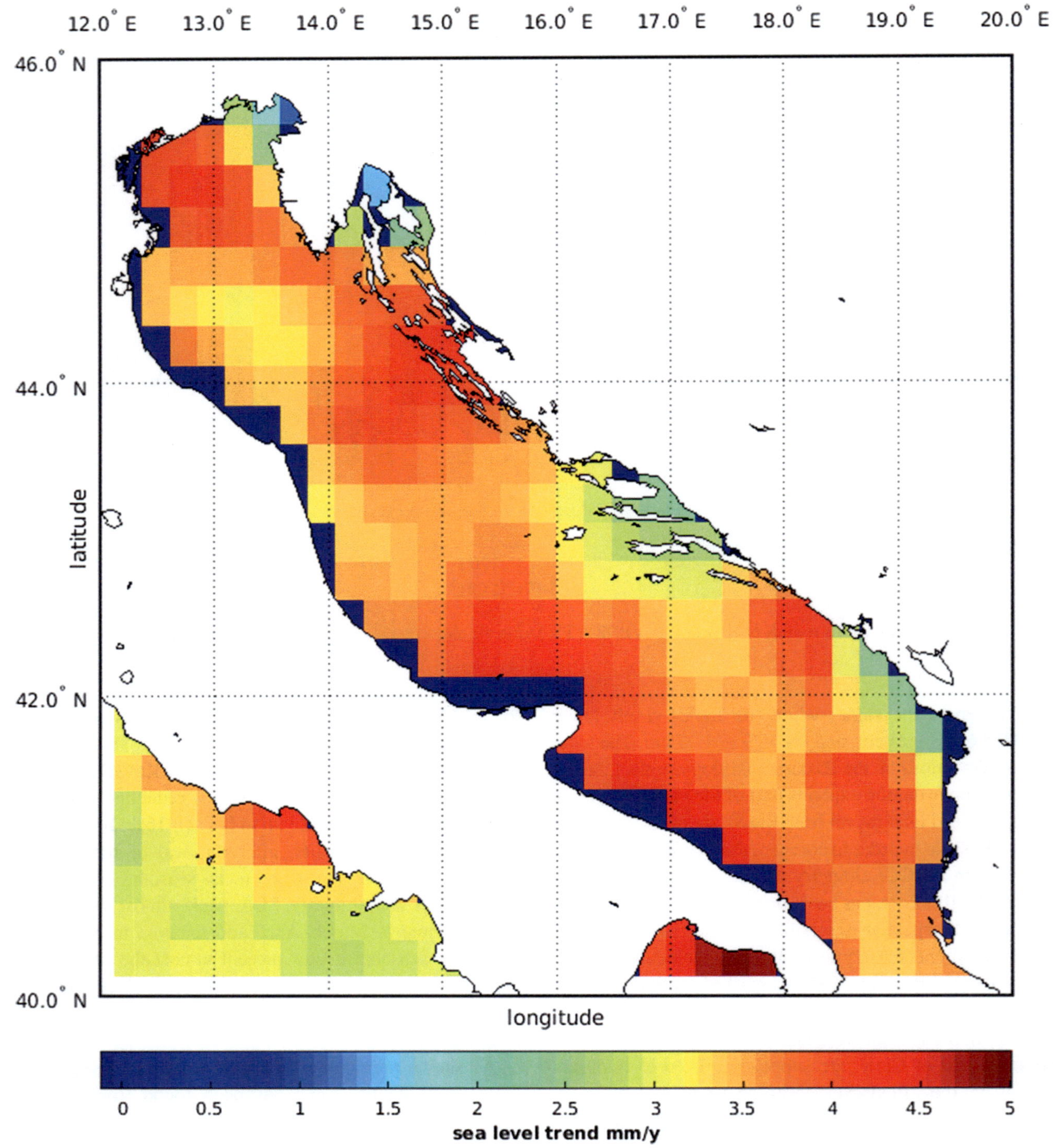

Fig. 5 Map of sea level trends in the Adriatic Sea from the CCI product (1993–2015)

1.40 mm year^{-1}: AAPTF rate is 5.65 ± 1.25 mm year^{-1} (IB corrected), ALT is 4.25 ± 1.25 mm year^{-1}. Baldin and Crosato (2017) conducted an independent study about the 2010–2015 vertical displacement of some local benchmarks in Venice, by comparing GPS data and other data derived from different methods in other studies. They report, for the period 2010–2015, a mean slope of about 1.45 mm year^{-1} (no std. reported). Specifically for the PS benchmark they found a vertical displacement rate of −1.46 ± 0.09 mm year^{-1} (Baldin 2018, personal communication). If this last rate could be assumed representing the whole period 1993–2015 for the Venice area, the TGs budget could be closed very easily: the difference TG–SA mutually cancels sea level variations

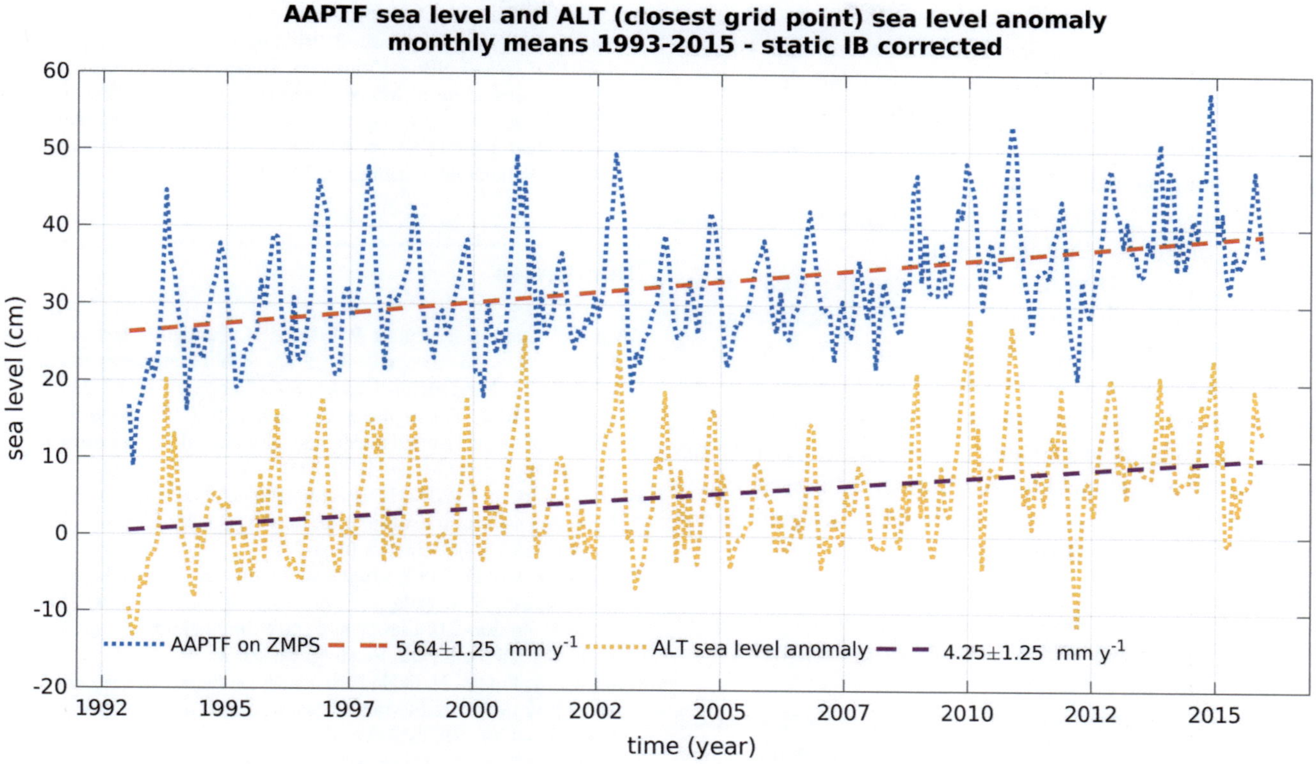

Fig. 6 Monthly means records of AAPTF SL and altimeter SLA (closest grid point). References are ZMPS for the AAPTF, and MSS for SLA

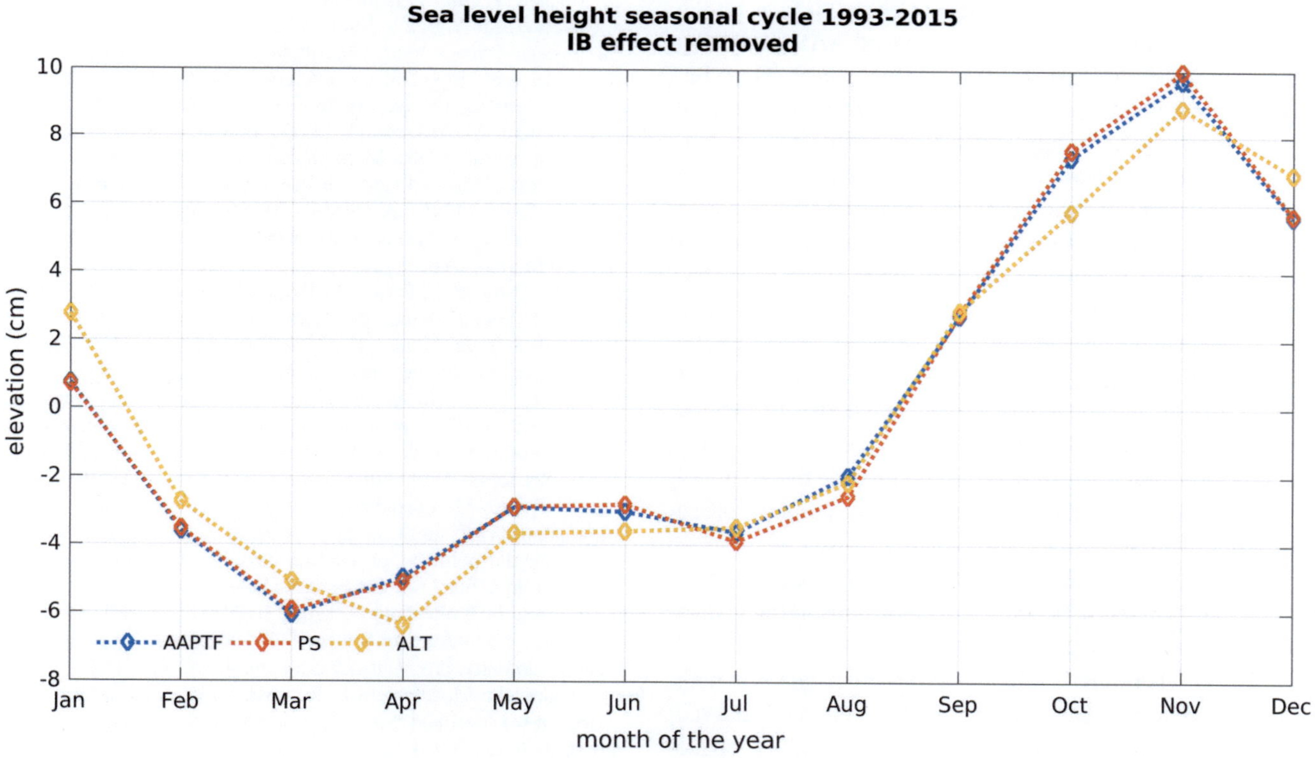

Fig. 7 Mean seasonal cycles of AAPTF and PS SL and altimeter SLA (closest grid point) 1993–2015

month by month (TG and SA observe the same SL, apart differences in the measurement procedure and other specific local sea level variations), except the crustal movement at the TG station. Indeed, calculating the residuals TG–SA for each month and then the trend of the residuals, for the AAPTF TG we obtain an estimate of the vertical land movement trend (1993–2015) of -1.40 ± 0.70 mm year^{-1}, very near to that found by Baldin and Crosato (2017).

Many other studies have investigated the VLM in the Venice area, over different periods, showing high spatial and temporal variations (e.g., Tosi et al. 2002). In addition to VLM, other aspects need to be further investigated, e.g., de-seasoning and serial auto-correlation. Also the removal of IB effect, including error estimation, deserves further investigations, as the derivation of the SL pressure difference used in this study mixes local atmospheric pressure observations and model reanalysis calculations.

4 Discussion and Concluding Remarks

In this paper, we analysed hourly sea level observations from three tide gauges around Venice in the northern Adriatic Sea. Monthly values have been derived from hourly values and compared with the altimetry dataset from the ESA Sea Level CCI. The comparison spans 23 years (1993–2015) using the nearest CCI grid point to the location of AAPTF, during which the trends are also estimated. Our results show that hourly observations computed from tide gauge match well while their trends, slightly differ from altimetry data set, because of local VLM. However, the coastal zone needs special treatment, and the increased uncertainty in this area is not reflected in the trend error in the present version of the Sea Level CCI, which is based on open ocean altimetry. During the CCI+ phase (2018–2019) the objective is to extend the satellite-based sea level climate record to the coastal zone with quality comparable to the open ocean. At the same time, the focus will be expanded also to include the VLM signal identification in the TG time series as this quantity, if accurately estimated, could provide the final tool in closing the balance between in situ- and altimetry-based estimate of SL rise.

Coastal altimetry has demonstrated that if standard products are reprocessed with dedicated algorithms, reliable data can be obtained up to few kms from the coasts (Vignudelli et al. 2011). Therefore, additional along-track data sets with consistent coastal processing for all missions and derived products dedicated to coastal regimes will be used to evaluate their current capabilities and perspectives for usage in long term sea level research studies.

Acknowledgements The authors want to thank the European Space Agency that funded the Climate Change Initiative to produce a climate quality record of sea level from satellite altimetry. This work is funded by ESA under Phase 2 Bridging Phase to CCI+ (contract n. 4000109872/13/I-NB – Contract Change Notice 6). The Centro Previsioni e Segnalazioni Maree (CPSM) of the Venice Municipality is acknowledged for providing the tide gauges in situ data and other meteorological parameters, under the CNR-ISAC/CPSM technical-scientific collaboration agreement n.0004298 13/10/2017.

References

Ablain M, Legeais JF, Prandi P, Marcos M, Fenoglio-Marc L, Dieng HB, Benveniste J, Cazenave A (2017) Satellite altimetry-based sea level at global and regional scales. Surv Geophys 38(1):7–31

Bajo M, De Biasio F, Umgiesser G, Vignudelli S, Zecchetto S (2017) Impact of using scatterometer and altimeter data on storm surge forecasting. Ocean Model 113:85–94

Baldin G, Crosato F (2017) L'innalzamento del livello medio del mare a Venezia: eustatismo e subsidenza. ISPRA, Quaderni Ricerca Marina, 10/2017. ISBN: 978-88-448-0861-7

Berrisford P, Dee D, Poli P, Brugge R, Fielding K, Fuentes M, Kallberg P, Kobayashi S, Uppala S, Simmons A (2011) The ERA-Interim archive, Version 2.0, European Centre for Medium Range Weather Forecasts Shinfield Park, Reading, Berkshire

Bonaduce A, Pinardi N, Oddo P, Spada G, Larnicol G (2016) Sea-level variability in the Mediterranean Sea from altimetry and tide gauges. Clim Dyn 47(9–10):2851–2866

Church JA, Clark PU, Cazenave A, Gregory JM, Jevrejeva S, Levermann A, Merrifield MA, Milne GA, Nerem RS, Nunn PD, Payne AJ, Pfeffer WT, Stammer D, Unnikrishnan AS (2013) Sea level change. In: Stocker TF, Qin D, Plattner G-K, Tignor M, Allen SK, Boschung J, Nauels A, Xia Y, Bex V, Midgley PM (eds) Climate change 2013: the physical science basis. Contribution of working group I to the fifth assessment report of the intergovernmental panel on climate change. Cambridge University Press, Cambridge, pp 1139–1216

Collini V, Ferla M, Trincardi F (2017) Previsioni delle Altezze di Marea per il Bacino San Marco e delle Velocità di Corrente per il Canal Porto di Lido – Laguna di Venezia Valori Astronomici 2017. COMUNE DI VENEZIA CPSM – ISPRA Istituto Superiore per la Protezione e la Ricerca Ambientale – CNR ISMAR Istituto di Scienze Marine, 76 pp

Cordella M, Zampato L, Pastore F, Tomasin A, Canestrelli P, Ferla M (2010) Le Tavole Annuali di Marea per Venezia. Atti dell'Istituto Veneto di Scienze, Lettere ed ARTI, Tomo CLXIX (2010–2011) – Classe di scienze fisiche, matematiche e naturali, pp 43–59

De Biasio F, Vignudelli S, Della Valle A, Umgiesser G, Bajo M, Zecchetto S (2016) Exploiting the potential of satellite microwave remote sensing to hindcast the storm surge in the Gulf of Venice. IEEE J Sel Topics Appl Earth Observ Remote Sens 9(11):5089–5105

De Biasio F, Bajo M, Vignudelli S, Umgiesser G, Zecchetto S (2017) Improvements of storm surge forecasting in the Gulf of Venice exploiting the potential of satellite data: the ESA DUE eSurge-Venice project. Eur J Remote Sens 50(1):428–441

Fenoglio-Marc L, Braitenberg C, Tunini L (2012) Sea level variability and trends in the Adriatic Sea in 1993–2008 from tide gauges and satellite altimetry. Phys Chem Earth Parts A/B/C 40:47–58

Ferla M, Cordella M, Michielli L, Rusconi A (2007) Long-term variations on sea level and tidal regime in the lagoon of Venice. Estuar Coast Shelf Sci 75:214–222

Legeais JF, Ablain M, Zawadzki L, Zuo H, Johannessen JA, Scharffenberg MG, Fenoglio-Marc L, Fernandes J, Baltazar Andersen O, Rudenko S, Cipollini P, Graham D, Quartly GD, Passaro M, Cazenave A, Cipollini P (2018) An improved and homogeneous altimeter sea level record from the ESA climate change initiative. Earth Syst Sci Data 10(1):281

Nerem RS, Beckley BD, Fasullo JT, Hamlington BD, Masters D, Mitchum GT (2018) Climate-change–driven accelerated sea-level rise detected in the altimeter era. Proc Natl Acad Sci U S A 115(9):2022–2025. https://doi.org/10.1073/pnas.1717312115

Orlić M (2015) The first attempt at cataloguing tsunami-like waves of meteorological origin in Croatian coastal waters. Acta Adriat 56(1):83–96

Quartly GD, Legeais JF, Ablain M, Zawadzki L, Fernandes MJ, Rudenko S, Carrère L, García PN, Cipollini P, Andersen OB, Poisson JC, Mbajon Njiche S, Cazenave A, Poisson JC (2017) A new phase in the production of quality-controlled sea level data. Earth Syst Sci Data 9(2):557

Rocco FV (2015) Sea level trends in the Mediterranean from tide gauges and satellite altimetry. Doctoral dissertation, Alma Mater Studiorum, University of Bologna, 128 pp

Scharroo R (2002) A decade of ERS satellite orbits and altimetry. PhD thesis, Delft University of Technology, Delft, 195 pp

Shirahata K, Yoshimoto S, Tsuchihara T, Ishida S (2016) Digital filters to eliminate or separate tidal components in groundwater observation time-series data. Jpn Agric Res Q 50(3):241–252

Tosi L, Carbognin L, Teatini P, Strozzi T, Wegmuller U (2002) Evidence of the present relative land stability of Venice, Italy, from land, sea, and space observations. Geophys Res Lett 29(12):1562

Vignudelli S, Kostianoy AG, Cipollini P, Benveniste J (eds) (2011) Coastal altimetry. Springer, Berlin, 578 pp. https://doi.org/10.1007/978-3-642-12796-0

Von Storch H, Zwiers FW (2001) Statistical analysis in climate research. Cambridge University Press, Cambridge, 496 pp

Wunsch C, Stammer D (1997) Atmospheric loading and the oceanic "inverted barometer" effect. Rev Geophys 35(1):79–107

Sentinel-3A: Validation of Orbit Products at the Copernicus POD Service

Jaime Fernández, Heike Peter, Emilio José Calero, Javier Berzosa, Luis Javier Gallardo, and Pierre Féménias

Abstract

The Copernicus POD (Precise Orbit Determination) Service is part of the Copernicus PDGS Ground Segment of the Sentinel-1, -2, and -3 missions. It is responsible of generating precise orbital products and auxiliary data files for their use as part of the respective PDGS processing chains.

For Sentinel-3, the CPOD (Copernicus Precise Orbit Determination) Service generates three types of products: the official Near Real Time (NRT) product, the Short Time Critical (STC) and the Non-Time Critical (NTC). The STC and NTC orbit products are generated as backup of the primary orbit products from CNES (Centre National d'Etudes Spatiales). The accuracy requirements (in the radial direction) are 10, 4 and 3 cm, respectively, but with the goal of achieving 8, 3 and 2 cm, respectively. Actual accuracies for Sentinel-3A are significantly better than the requirements and goals, as it will be shown in this paper.

Considering the importance of the Sentinel-3 orbit products for the radar altimetry processing the orbit validation is crucial. The validation of the different Sentinel-3 orbital products from the Copernicus POD Service, therefore, consists of several independent steps including orbit overlap analysis, direct orbit comparison, but also cross-validation with SLR and DORIS measurements. The different orbit validation steps are described and results are shown for the entire mission time until March 2018 of Sentinel-3A.

Keywords

Copernicus POD Service · Orbit validation · Sentinel-3A

1 Introduction

Sentinel-3A is the first satellite of the Copernicus Sentinel-3 mission launched on February 16, 2016. By now, more than 2 years of data are available for many scientific applications including sea level height measurements provided by the radar altimeter. The altimeter observations from Sentinel-3A contribute to the long time series of observations from previous and current altimeter satellites like TOPEX/Poseidon, Jason-1, -2, -3, Saral/Altika, and HY2-A (incomplete list). The availability of precise orbits is crucial for the analysis of altimeter observations and in particular the long-term stability of the orbit accuracy. Validation of the orbital products is, therefore, essential. In the frame of Sentinel-3 mission, the

J. Fernández · E. J. Calero · J. Berzosa · L. J. Gallardo
GMV AD, Tres Cantos, Spain
e-mail: jfernandez@gmv.com; ecalero@gmv.com; jberzosa@gmv.com; ljgallardo@gmv.com

H. Peter (✉)
PosiTim UG, Seeheim-Jugenheim, Germany
e-mail: heike.peter@positim.com

P. Féménias
ESA/ESRIN, Frascati, Italy
e-mail: pierre.femenias@esa.int

S. P. Mertikas, R. Pail (eds.), *Fiducial Reference Measurements for Altimetry*,
International Association of Geodesy Symposia 150, https://doi.org/10.1007/1345_2019_64

accuracy checks of the orbital products are done routinely: on daily basis, the two operational solutions, from CNES and Copernicus POD Service (CPOD) are compared, in addition to an external comparison against an ESOC solution. These metrics are published on daily quality control reports for internal monitoring. Then quarterly several independent centres (AIUB, DLR, TUM, TU Delft, EUMETSAT, CLS) in addition to CNES, CPOD and ESOC, generate orbital solutions, which are used to generate a complete cross-validation in order to assess the performance during the last 4 months. These reports are currently being published on the Sentinels OnLine webpage (https://sentinel.esa.int/web/sentinel/missions/sentinel-3/ground-segment/pod/documentation).

Different orbital products for Sentinel-3A are provided by the Copernicus POD Service, with differences in term of latency and accuracy requirements. The validation of the orbital accuracy is based on several indices: from the comparison against independently derived orbits, to the validation with SLR (Satellite Laser Ranging) measurements or with DORIS-derived orbits. These various tasks are needed to guarantee the required orbit accuracy of 2–3 cm in radial direction not only on short-term but also on long-term.

First analyses of the Sentinel-3A orbit results from CPOD were already presented in Fernández et al. (2016, 2017a), from ESOC in Otten et al. (2016) and from CNES in Couhert et al. (2016). These analyses focussed on shorter time intervals and were not going into details as done in the following. The analysis presented here is based on the data from the beginning of the operational mission (partly also commissioning phase data) until end of January 2018. Investigations on the long-term stability are only done as a preliminary analysis, future work has to follow in this perspective.

In the following sections firstly the Copernicus POD Service is described with focus on the Sentinel-3 mission. The different methods to validate the orbit accuracy are outlined and applied to the Sentinel-3A orbital products. Future steps to improve the accuracy are discussed as well.

2 The Copernicus POD Service: Sentinel-3

The Copernicus POD Service (Fernández et al. 2015) is a European consortium led by GMV, Spain, consisting of the several members from all over Europe.

The service is responsible for providing orbital and auxiliary products for the Sentinel-1, -2, and -3 missions. The products are delivered to the Copernicus PDGS of the individual missions. They are used to generate core products for the missions, like SAR/InSAR products of Sentinel-1, or Surface Topography Mission (STM) products of Sentinel-3 (currently CNES orbital products are used operationally for the S3 STC and NTC STM products, CPOD solution is the back-up).

Currently, the CPOD Service operationally processes five satellites, the A- and B-satellites from Sentinel-1 and -2, and the A-satellite from Sentinel-3. Sentinel-3B has been launched on 25th of April 2018 and will be operational after 4–5 months commissioning phase. All satellites need orbit products with different latency and accuracy requirements.

The CPOD Service is supported by the Copernicus POD QWG (Quality Working Group). The QWG is an integral part of the CPOD Service and the core members are institutions with a long LEO (Low Earth Orbit) POD expertise, namely AIUB (e.g., Jäggi et al. 2006; Bock et al. 2014), DLR (e.g., Montenbruck et al. 2008), TUD (e.g., Visser et al. 2009; van den IJssel et al. 2015), TUM (Švehla and Rothacher 2003) and ESOC (e.g., Flohrer et al. 2011). Independent orbit solutions delivered by this group are used to validate the CPOD results and the recommendations from this body guarantee that the CPOD Service follows state-of-the-art algorithms, models, and conventions.

The software core used by the CPOD Service is NAPEOS (Navigation Package for Earth Orbiting Satellites, Springer et al. 2011). Although the characteristics and the requirements are different for the three missions, the same core POD setup is used to the largest extent possible. This strategy facilitates maintenance of the complex system of the CPOD Service. The POD instruments for Sentinel-3 are two (redundant) GPS receivers, one DORIS receiver and one laser retro reflector (LRR) allowing for SLR tracking to the satellite. The Sentinel-3A orbit products from the CPOD Service are all based on the GPS observations only, the DORIS observations are not used. The SLR measurements are also not used for the orbit determination but for external validation of the GPS-based orbit products. Table 1 summarises the latency and accuracy requirements of the NRT (near real-time), STC (short-time critical), and NTC (non-time critical) orbit products delivered by the CPOD Service for Sentinel-3.

STC and NTC products are computed in the premises of the Copernicus POD Service in GMV (Tres Cantos – Madrid), while the NRT products are computed directly on the Marine and Land PDGS (currently located in EUMETSAT and Svalbard, respectively) by the so-called S3PODIPF, an Instrument Processing Facility (IPF) developed and maintained by GMV; while it is executed on the PDGS,

Table 1 Latency and accuracy requirements for orbital products from Sentinel-3

Category	Latency	Radial accuracy (RMS)
NRT	30 min	10 cm (target: 8 cm)
STC	1.5 days	4 cm (3 cm)
NTC	25 days	3 cm (2 cm)

GMV is still responsible for the latency and quality of the product.

3 Orbit Accuracy Validation Methods for Sentinel-3A

3.1 Processing Metrics

The following metrics of the processing are a first measure for the performance of the orbit solutions:

- Carrier phase RMS
- Number of observations used
- Number of rejected observations
- Presence of data gaps
- Presence of manoeuvre

The carrier phase RMS is very consistent for the entire mission and has a mean value of 6.3 mm. The percentage of observations used is at 98.84% and only on days with manoeuvres this percentage might be lower.

3.2 Orbit Overlap Analysis

The arc length for the orbital solutions is longer than the arc length required for the specific product line. This is done to avoid boundary effects at the beginning and end of the arcs. The orbit overlap analysis is in particular feasible for the NTC orbits. It gives information about the consistency of the orbit solutions.

The mean radial and 3D RMS of 4 h overlaps of the CPOD NTC orbits centred around midnight are 0.27 cm and 1.05 cm, respectively. The values are very stable although the first months of the mission (commissioning phase) show slightly larger variations.

3.3 Comparison to Other Orbit Solutions

Comparison to orbit solutions from other product lines or from other institutions is a common tool to assess the orbit quality. Orbits from other institutions are in most cases generated with different software tools, partly based on different background models such as the Earth's gravity field or the Ocean tide model. The non-gravitational force modelling is different or even done completely empirical. Details on the parameters and models used for the reduced-dynamic orbit determination at the CPOD Service (also for all other QWG solutions) can be found in the 4-monthly reports from the Regular Service Reviews (https://sentinel.esa.int/web/sentinel/missions/sentinel-3/ground-segment/pod/documentation). For comparison the standards used for the CNES orbit solutions are available at ftp://cddis.gsfc.nasa.gov/pub/reports/slrcnes/POD_configuration_GDRE.pdf .

The comparison allows validation of the approaches against each other. The consistency of the orbit solutions can be checked. The comparison between different orbit product lines is at first used to assess the orbit quality for the products with shorter latency (NRT, STC), which have weaker accuracy requirements than the NTC orbit product.

The CPOD Service is not the only provider of operational S-3A orbit solutions. Figure 1 shows the manifold number of orbit solutions being available for cross-comparison. In addition to the CPOD Service (red boxes) also CNES (blue boxes) is providing operational orbit solutions for Sentinel-3A. The arrows indicate the comparisons between the different products.

Table 2 lists the mean radial and 3D RMS values of comparisons between different NRT and STC orbit solutions. The statistics have to be looked at as an ensemble. They show the improvement of the orbit quality from the on-board navigation solution (NAVSOL) to the MOEORB (CPOD STC solution).

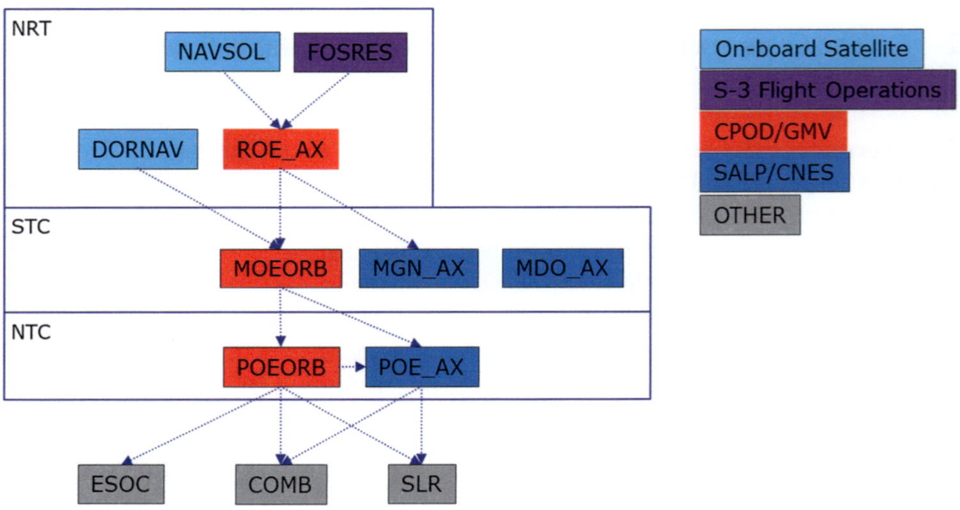

Fig. 1 Orbit validation scheme for Sentinel-3A orbit products

Table 2 Mean radial and 3D RMS of different orbit comparisons (NRT (*underlined*) and STC)

Orbit comparison		Mean radial RMS ± std. dev. (cm)	Mean 3D RMS ± std. dev. (cm)
FOS	ROE_AX	18.5 ± 10.5	101.8 ± 24.6
NAVSOL	ROE_AX	52.9 ± 16.5	89.8 ± 21.3
DORISNAV	MOEORB	3.2 ± 1.2	10.8 ± 3.3
ROE_AX	MOEORB	1.0 ± 0.8	3.4 ± 1.6

Table 3 Mean radial RMS of different orbit comparisons (STC and NTC (**bold**))

Orbit comparison		Mean radial RMS ± std. dev. (cm)
MOEORB	POEORB	0.6 ± 0.5
MOEORB	POE_AX	0.9 ± 0.5
MGN	POEORB	0.7 ± 0.5
MGN	POE_AX	0.5 ± 0.3
MDO	POEORB	0.9 ± .0.4
MDO	POE_AX	0.7 ± 0.2
ESOC	POEORB	**0.7 ± 0.6**
POE_AX	POEORB	**0.6 ± 0.1**
POEORB	COMB	**0.6 ± 0.1**
POE_AX	COMB	**0.6 ± 0.1**

Table 3 lists mean radial RMS values of comparisons between STC and NTC orbits. The STC orbit MOEORB is already very close to the NTC orbits with mean radial RMS values of 0.6 cm (POEORB) and 0.9 cm (POE_AX). The lower value for the comparison to POEORB can be related to the fact that MOEORB and POEORB are generated with the same software and the same background models. The two CNES STC orbits MGN (GPS) and MDO (DORIS) compare just as well to the CNES NTC (POE_AX) as to the CPOD NTC (POEORB) orbit product. The orbit comparisons show the high consistency between the different orbit solutions on the few mm level in radial direction.

Figure 2 shows the radial RMS for comparisons of the CPOD NTC POEORB and the CNES NTC orbits to a combined orbit product, respectively. The mean radial RMS is 0.6 cm for both NTC orbit products. The combined orbit product is a weighted average of several orbit solutions provided by the Copernicus POD QWG. The combination procedure is the same as used for the combined GPS orbit products of the IGS (International GNSS Service, Beutler et al. 1995; Dow et al. 2009; Johnston et al. 2017). Drifts in the radial RMS from CPOD and CNES with respect to the combined orbit are very small with −0.62 mm/y and +0.52 mm/y, respectively. Nevertheless, the drifts are in opposite direction and the development of the drifts has to be further monitored and checked.

In addition to a chronological comparison of the time series a geographical comparison is done. The mean radial differences between NTC CPOD POE and CNES orbits for the time span from June 2016 to March 2018 are shown in a geographical distribution in Fig. 3 (top). Figure 3 (middle) further shows the mean radial errors of CNES orbits compared to the combined QWG orbit and of CPOD POE orbits compared to the combined QWG orbit (Fig. 3, bottom). Systematic radial orbit differences are visible between the products. The cause of these differences has to be further analysed to also assure a long-term stability of the orbits.. It has to be investigated if the different gravity field model (EIGEN6S instead of EIGEN.GRGS.RL03.v2) used for the first months (until 12th December 2016ˆ) for the CPOD NTC solutions might be responsible for the systematic orbit differences or if differences in the radiation pressure modelling (e.g., CNES with re-radiation, CPOD without re-radiation) causes these systematics as already shown for Jason-1 (Zelensky et al. 2010).

3.4 External Validation

The external validation can be divided into two categories. The first one is the validation based on measurements or

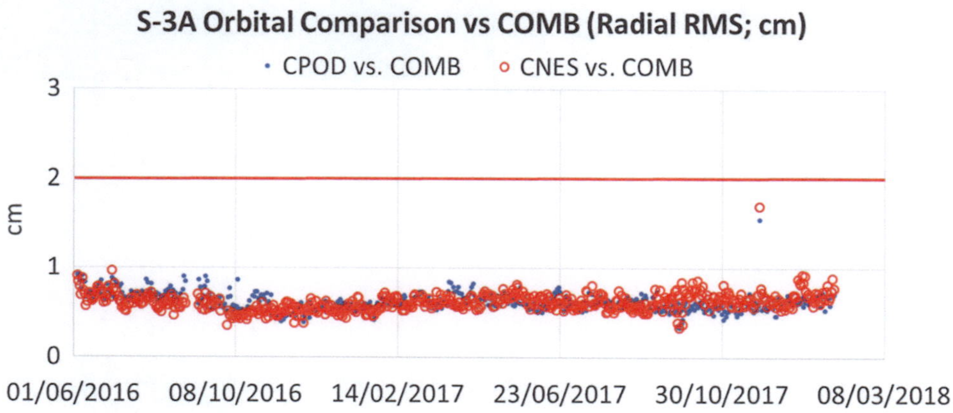

Fig. 2 Radial RMS (cm) of orbit comparison between CPOD POE (NTC) and CNES NTC versus a combined orbit product

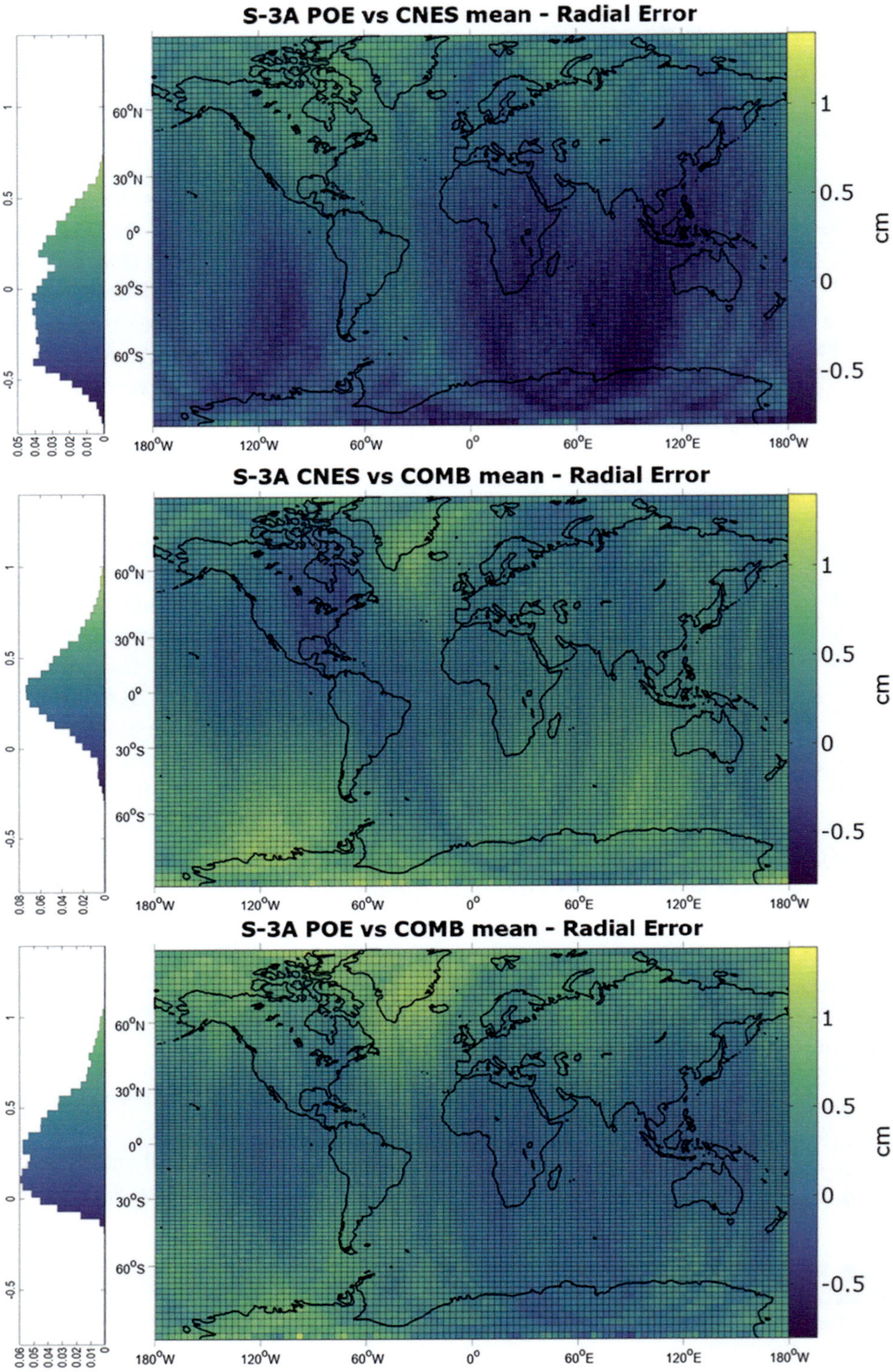

Fig. 3 Mean radial error (cm) between S-3A POE (CPOD NTC) and CNES NTC orbit (top), CNES NTC and combined orbit (middle), POE and combined orbit (bottom); June 2016–March 2018. On the left, the histogram of the residuals is displayed

orbits from other POD techniques such as SLR residual analysis or comparison to DORIS-derived orbits (not shown here). The second category is the validation based on other measurements such as altimeter cross-over analysis (Shum et al. 1990) (also not shown here).

The SLR validation for the Sentinel-3A satellite is done for the NTC orbits. A subset of nine stations is selected to perform the validation (Fernández et al. 2017b) on the basis of stations delivering highest level measurement to the satellite. The performance of all stations tracking Sentinel-3A (and in future Sentinel-3B) is regularly monitored. A yearly report (Gallardo 2018) summarising the performance is provided to the ILRS (International Laser Ranging Service).

Figure 4 shows the mean + standard deviation (left) and the RMS (right) of SLR residuals for the CNES orbit (top) and CPOD POE orbit (bottom) validation depending on the elevation of the SLR observations. Additionally, the number of observations used for the validation is plotted (red). The excellence performance of both orbit products can be seen with mean SLR residuals smaller than 0.5 cm and RMS values of smaller than 2 cm for both orbits even when considering the observations at all elevations.

It can also clearly be seen that the number of observations is very low when cutting the observations already at high elevation angles. From zenith to 70° elevation only 2.18% of the observations are available. This cut-off angle is often used for the validation of altimeter satellites.

Figure 5 shows in the left column the mean of the SLR residuals in a geographical distribution when a cut-off angle of 0° is used. The top panel shows the results for the CNES orbits and the bottom panel for the CPOD POE orbits, respectively. The right column of Fig. 5 shows the corresponding results when applying a cut-off angle of 60°. Already in the left column of Fig. 5 it can be noted that the distribution of the SLR observations is not well spread over the globe. If cutting off the observations at 60° (Fig. 5, right column) the distribution is even sparser and only small parts

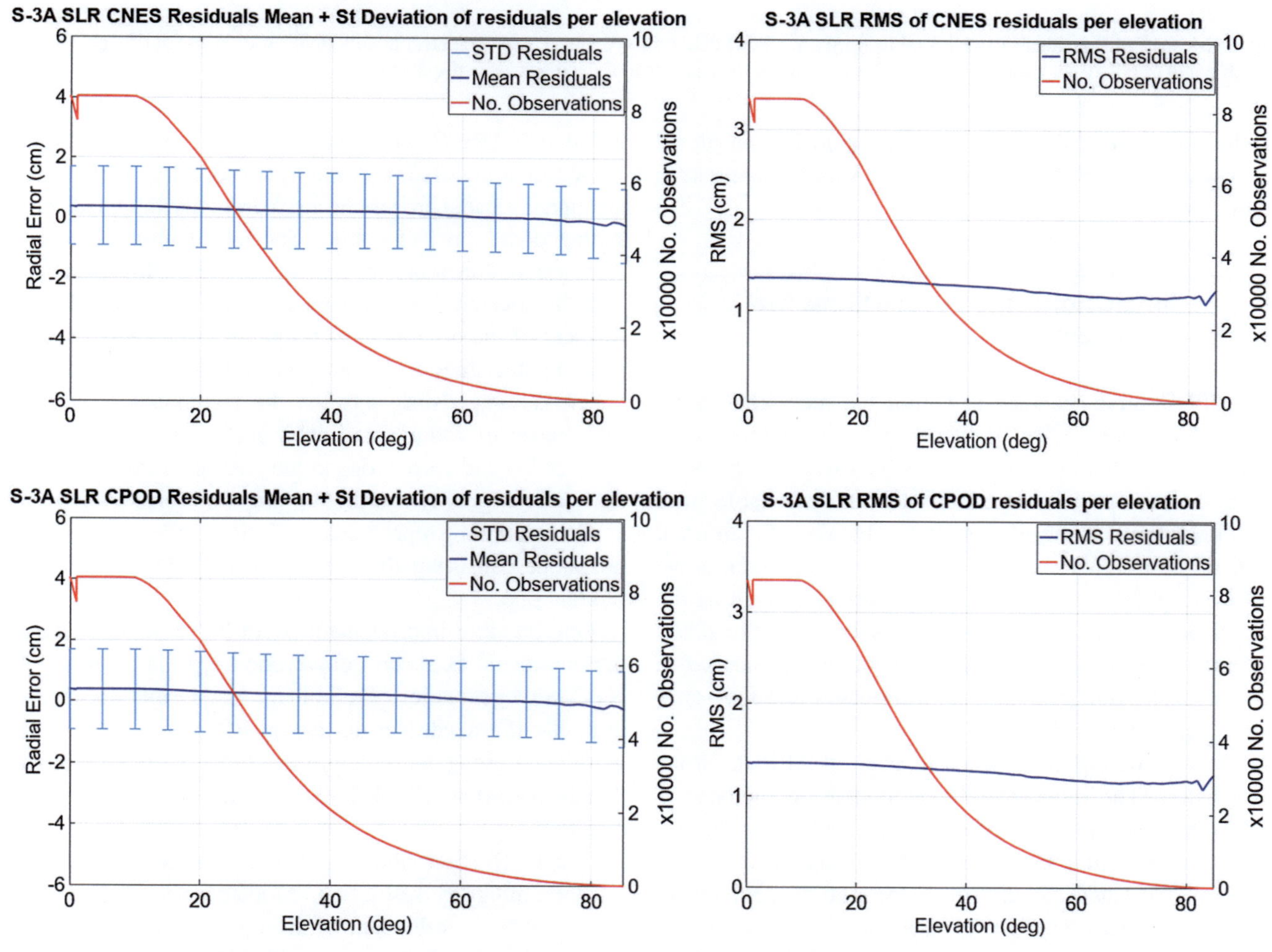

Fig. 4 Mean + Standard deviation of SLR residuals (left column) for validation of CNES orbits (top) and CPOD POE orbits (bottom) depending on elevation of SLR observations; corresponding RMS (cm) (right column)

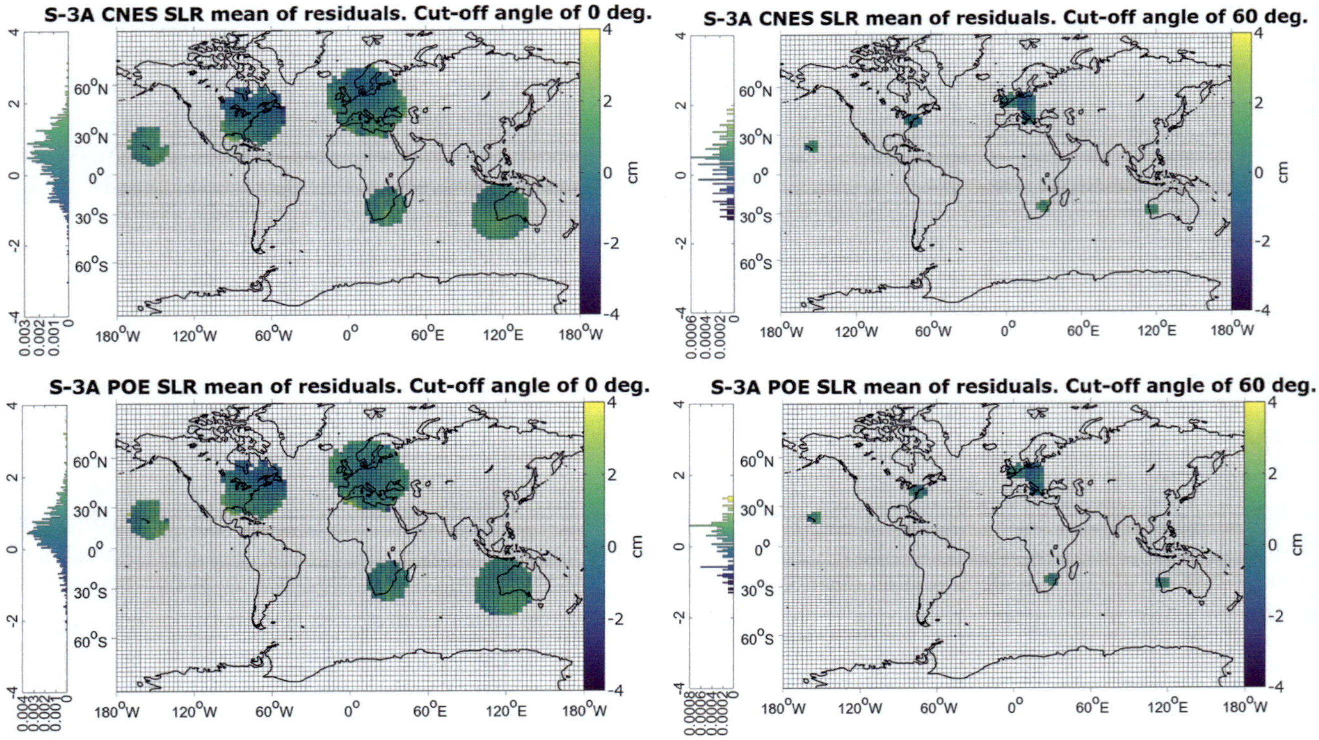

Fig. 5 Mean of SLR residuals to CNES orbits (top) and CPOD POE orbits (bottom) in geographical distribution; cut-off angle of 0° (left column); cut-off angle of 60° (right column). On the left of the maps, the histogram of the residuals is displayed

of the orbits are validated. This fact has to be considered when using the SLR statistics as external validation measures of the orbits.

4 Conclusions and Future Plans for Improvement

The CPOD orbit accuracy is within the requirements for all product lines. A lot of effort is taken to thoroughly validate the different orbit products, in particular the NTC orbit product. Orbit comparisons to other available orbit solutions show a very high consistency between the different orbit solutions. External validation with SLR confirms the excellent quality of the CPOD NTC orbits as well as the CNES orbit product. The differences between the orbits are very small and additional investigations are needed to analyse systematic geographical differences and long-term trends between the orbits.

Effort is, nevertheless, taken to make the CPOD orbits even better. The dynamical force models are updated if better and more recent models are available, e.g., instantaneous re-radiation has recently been implemented (3rd October 2018) and the update of the gravity field model (EIGEN-GRGS.RL03.v2) to a newer release is foreseen after corresponding validation. Updates of the IERS (International Earth Rotation Service) conventions (Petit and Luzum 2010) and ITRF (International Terrestrial Reference Frame, Altamimi et al. 2016) are also followed. All model changes and updates in the processing are thoroughly validated before setting operational of course. Special attention is taken if software modifications are needed for the model updates.

Sophisticated analysis is on-going to investigate the correctness of the used antenna reference point (ARP) coordinates for the three techniques (GPS. DORIS, SLR). Based on cross-comparisons between the techniques, systematic differences are found (Ait-Lakbir et al. 2017; Montenbruck et al. 2018) and corrections to the ARP coordinates or even the center of gravity coordinates might be considered.

Corresponding reprocessing efforts are taken of course to guarantee consistent time series for the CPOD NTC orbit product line.

One possible improvement could be to include DORIS measurements as second observation type into the CPOD NTC orbit generation. Preliminary test results are already available (Fernández et al. 2017a; Peter et al. 2018). The work is on-going and finally it has to be thoroughly proved that the inclusion of DORIS improves the orbit determination results.

The GPS observation modelling can be improved by integer ambiguity resolution. Montenbruck et al. (2018) showed that the orbit quality is improved when fixing the phase ambiguities following the CLS/GRGS approach (Laurichesse et al. 2009). Other ambiguity resolution approaches are also possible for the LEO case, e.g., including the LEO into a ground network for ambiguity resolution. It has to be verified in the future which is the most suitable method for Sentinel-3 at the CPOD Service.

The orbit validation itself could further be improved by adding routinely validation with other techniques such as altimeter cross-over analyses.

Acknowledgements The Copernicus POD Service is financed under ESA contract No. 4000108273/13/1-NB. The work performed in the frame of this contract is carried out with funding by the European Union. The views expressed herein can in no way be taken to reflect the official opinion of either the European Union or the European Space Agency. The authors thank the anonymous reviewer for the constructive comments.

References

Ait-Lakbir H, Couhert A, Mercier F, Houry A, Jalabert E, Moyard J (2017) Impact of the next foreseen IERS mean pole model (linear) on altimeter satellite precise orbits, and validation of updated measurement models. In: Poster at OSTST 2017, Miami, FL, USA

Altamimi Z, Rebischung P, Métivier L, Collilieux X (2016) ITRF2014: a new release of the International Terrestrial Reference Frame modeling nonlinear station motions. J Geophys Res Solid Earth 121:6109–6131. https://doi.org/10.1002/2016JB013098

Beutler G, Kouba J, Springer T (1995) Combining the orbits of the IGS analysis centres. Bull Geod 69:200. https://doi.org/10.1007/BF00806733

Bock H, Jäggi A, Beutler G et al (2014) GOCE: precise orbit determination for the entire mission. J Geod 88(11):1047–1060. https://doi.org/10.1007/s00190-014-0742-8

Couhert A, Mercier F, Jalabert E, Moyard J, Houry S, Ait-Lakbir H (2016) Status of the CNES precise orbit ephemerides for Sentinel-3A. In: Poster at OSTST 2016, November 2016, La Rochelle, France

Dow JM, Neilan RE, Rizos C (2009) The International GNSS Service in a changing landscape of Global Navigation Satellite Systems. J Geod 83:191–198. https://doi.org/10.1007/s00190-008-0300-3

Fernández J, Escobar D, Ayuga F et al (2015) Copernicus POD service operations. In: Proceedings of the Sentinel-3 for science workshop, 2–5 June 2015, Venice, Italy

Fernández J, Peter H, Fernández C, Féménias P, Labroue S, Ollivier A (2016) Sentinel-3 precise orbit determination at the Copernicus POD Service. In: Poster at OSTST 2016, November 2016, La Rochelle, France

Fernández J, Fernández C, Peter H, Féménias P (2017a) Sentinel-3 precise orbit determination at the Copernicus POD Service. In: Poster at OSTST 2017, October 2017, Miami, Florida, USA

Fernández J, Fernández C, Calero EJ, Gallardo LJ, Peter H, Féménias P (2017b) The Copernicus Sentinel-3 mission. In: Presented at the 2017 ILRS Technical Workshop, Riga, Latvia, October 2–5, 2017

Flohrer C, Otten M, Springer T et al (2011) Generating precise and homogeneous orbits for Jason-1 and Jason-2. Adv Space Res 48(1):152–172. https://doi.org/10.1016/j.asr.2011.02.017

Gallardo LJ (2018) Sentinel-3 SLR yearly report – 2017. https://ilrs.cddis.eosdis.nasa.gov/docs/2017/GMV-GMESPOD-SLR-0002_v1.1_Sentinel-3_SLR_Yearly_Report-2017.pdf

Jäggi A, Hugentobler U, Beutler G (2006) Pseudo-stochastic orbit modeling techniques for low-Earth orbiters. J Geod 80(1):47–60. https://doi.org/10.1007/s00190-006-0029-9

Johnston G, Riddell A, Hausler G (2017) The international GNSS service. In: Teunissen PJG, Montenbruck O (eds) Springer handbook of global navigation satellite systems, 1st edn. Springer, Cham, pp 967–982. https://doi.org/10.1007/978-3-319-42928-1

Laurichesse D, Mercier F, Berthias JP, Broca P, Cerri L (2009) Integer ambiguity resolution on undifferenced GPS phase measurements and its application to PPP and satellite precise orbit determination. Navigation 56(2):135–149. https://doi.org/10.1002/j.2161-4296.2009.tb01750.x

Montenbruck O, Andres Y, Bock H et al (2008) Tracking and orbit determination performance of the GRAS instrument on MetOp-A. GPS Solut 12(4):289–299. https://doi.org/10.1007/s10291-008-0091-2

Montenbruck O, Hackel S, Jäggi A (2018) Precise orbit determination of the Sentinel-3A altimetry satellite using ambiguity-fixed GPS carrier-phase observations. J Geod 92:711. https://doi.org/10.1007/s00190-017-1090-2

Otten M, Boomkamp H, Springer T, Enderle W (2016) Precise and homogeneous orbits for Sentinel-3A. In: Presentation at OSTST 2016, November 2016, La Rochelle, France

Peter H, Fernández J, Féménias P (2018) Copernicus POD Service – Sentinel-3A orbit determination based on DORIS observations. In: Poster at 4th Sentinel-3 validation team meeting, March 13–15, 2018, EUMETSAT, Darmstadt, Germany

Petit G, Luzum B (eds) (2010) IERS conventions 2010. IERS technical note 36. Verlag des Bundesamts für Kartographie und Geodäsie, Frankfurt

Shum CK, Zhang BH, Tapley BD, Schutz BE (1990) Altimeter crossover methods for precision orbit determination and the mapping of geophysical parameters. J Astronaut Sci 38(3):355–368

Springer T, Dilssner F, Escobar D (2011) NAPEOS: the ESA/ESOC tool for space geodesy

Švehla D, Rothacher M (2003) Kinematic and reduced-dynamic precise orbit determination of low Earth orbiters. Adv Geosci 1:47–56. https://doi.org/10.5914/adgeo-1-47-2003

van den IJssel J, Encarnação J, Doornbos E et al (2015) Precise science orbits for the Swarm satellite constellation. Adv Space Res 56(6):1042–1055. https://doi.org/10.1016/j.asr.2015.06.002

Visser P, van den IJssel J, van Helleputte T et al (2009) Orbit determination for the GOCE satellite. Adv Space Res 43(5):760–768. https://doi.org/10.1016/j.asr.2008.09.016

Zelensky N, Lemoine FG, Ziebart M et al (2010) DORIS/SLR POD modeling improvements for Jason-1 and Jason-2. Adv Space Res 46(12):1541–1558. https://doi.org/10.1016/j.asr.2010.05.008

The DTU17 Global Marine Gravity Field: First Validation Results

O. B. Andersen and P. Knudsen

Abstract

The most recent released global marine gravity field from DTU Space takes into account the new SARAL/AltiKa geodetic mission initiated in 2016 along with new improved Arctic processing of the Cryosat-2 mission. With its 369 days repeat cycle, Cryosat-2 provides one repeat of geodetic mission data with 8 km global resolution each year since its launch in 2010. Together with the Jason-1 end-of-life geodetic mission in 2012 and 2013, we now have more than five times as many geodetic missions sea surface height observations compared with the old ERS-1 and Geosat geodetic missions.

The DTU17 has been derived focusing on improving the coastal and Arctic gravity field, enhancing the shorter wavelength of the gravity field (10–15 km). For DTU17, we find a substantial improvement in marine gravity mapping as shown through comparison with high quality airborne data flown north of Greenland in 2009.

Keywords

Arctic Ocean · Free air gravity · Satellite altimetry

1 Gravity Field Update

Since the release of the DTU15 global marine gravity field in 2015 (Andersen et al. 2017) a number of additional data have become available to marine gravity field mapping. This means, that data from the first generation altimeters like ERS-1 and Geosat are now retired and not used anymore for marine gravity field modelling and only data from the new second generation altimeters are used.

Cryosat-2 continues to provide data along its 369 day near repeat since 2010 completing 7 full geodetic cycles for the derivation of DTU17. During the period from May 2012 to June 2013 the Jason-1 satellite operated in a 406 days geodetic mission as part its end of life mission. Jason-1 is particularly valuable for both global high resolution gravityfield modelling and bathymetry modelling at low to mid latitude (Sandwell et al. 2014). However, the 66° inclination of Jason-1 prevents it providing data at high latitude. Cryosat-2 has an inclination of 88° providing data throughout the Arctic Ocean up to 88°N or 220 km from the North Pole.

Since early 2016, a third geodetic mission by the SARAL/AltiKa satellite has accidently become available. Due to technical issues with the reaction wheels the operators decided to pursue this mission with a new phase named "SARAL-DOP" for SARAL-Drifting Orbit phase. By not maintaining the 35-day repetitive ground track the natural decay of the orbit creates a so-called uncontrolled geodetic orbit and provides data up to 82°N. Such uncontrolled orbit is similar to the way Geosat operated.

The SARAL/AltiKa operates at Ka-band at a pulse repetition frequency of 4,000 Hz were all altimetric satellites operates at Ku band, typically with a pulse repetitions frequency of 2,000 Hz. SARAL provides two important improvements to altimetric gravity field modelling. Firstly, the higher pulse repetition frequency generate higher number of ("independent") observations, which can be averaged to

O. B. Andersen (✉) · P. Knudsen
DTU Space, Copenhagen, Denmark
e-mail: oa@space.dtu.dk

S. P. Mertikas, R. Pail (eds.), *Fiducial Reference Measurements for Altimetry*,
International Association of Geodesy Symposia 150, https://doi.org/10.1007/1345_2019_65

lower the range precision. Secondly, the Ka-band altimeter has a significantly smaller footprint on the sea surface. The smaller footprint is particularly important for coastal and sea ice contaminated regions, as less sea surface height observations are corrupted by the presence of land or ice inside the footprint (Fu and Cazenave 2001). The footprint size is a function of the altitude of the spacecraft and the significant wave-height (Fu and Cazenave 2001; Chelton et al. 1989) and is given below for a 2-m wave-height. ERS-1/2, Envisat and Geosat all had a footprint around 70 km^2; Jason-1/2/3 have a footprint size of 95 km^2 as it flies at nearly twice the altitude. However, the SARAL Ka-band altimeter has a footprint size of 40 km^2.

For Cryosat-2 the footprint is similar to ERS-1 when the satellite operates in conventional or low resolution mode (LRM). However, when the satellite operates in SAR mode (Raney 1998), the footprint is sliced up in 300 m beams across the flight direction reducing the footprint to less than 5 km^2. Cryosat-2 occasionally operates in SARin where the secondary receiving antenna is activated. The Cryosat-2 mode mask (https://earth.esa.int/web/guest/-/geographical-mode-mask-7107) dictates which mode is active where. This mode mask is updated regularly to accommodate user request. The mode-mask is a consequence of bandwidth and the limited ability to transfer the high resolution SAR and SARin data to the ground.

The Arctic Ocean has been measured in SAR or SARin mode throughout the mission and, from an altimetric gravity point of view, the Cryosat-2 SAR and SARin data are processed identically.

SAR (and SARin) altimetry has a further advantage to conventional LRM altimetry enabling an improved recovery of shorter wavelength of the gravity field. This stems from the fact, that SAR altimetry does not suffer from correlated noise in the sea surface height observations in the 10–50 km wavelength seen for conventional altimetry (Dibarboure et al. 2014). Various ways have been employed to mitigate this for conventional altimetry (i.e., two-step retracking by Sandwell et al. 2013). However, this is not needed for SAR altimetry (Sandwell et al. 2014).

In order to exploit the full potential of Cryosat-2 SAR and SARin data, we have retracked the Cryosat-2 Level 1B waveform data from the recently updated Baseline-C using the narrow peak retracker (Jain et al. 2015). This empirical retracker is very robust and is able to provide accurate heights even when the waveform is moderately contaminated by the presence of sea ice (Stenseng and Andersen 2012). Hence, this retracker is able to retrieve sea level from leads which are significantly smaller than the footprint and which can be as small as 10–20 m.

The geodetic mission sea-surface heights are corrected for range corrections (Andersen and Scharroo 2011) and processed to extract the residual geoid information following the methods described in Andersen et al. (2010b). Similar remove-restore technique relative to EGM2008 (Pavlis et al. 2012) used for previous DTU marine gravity fields was applied, and the data were processed in global mesh of tiles of 1° by 3° latitude by longitude.

For DTU15 global marine gravity field, the Goddard Ocean tide (GOT4.8, Stammer et al. 2014) global ocean tide model was preferred. However, in some coastal regions, this model limits the data with ocean tide correction due to its coarse resolution of 0.5° compared with more recent ocean tide models like FES2014 (Carrere et al. 2015), which has a resolution of 0.125°. One of the regions where the coarse resolution of the GOT4.8 ocean tide model results in a large number of the Cryosat-2 observations being rejected, due to missing ocean tide correction is the south coast of Australia close to Adelaide. Figure 1 illustrates the problem around Adelaide. The number of valid Cryosat-2 observations with the GOT4.8 ocean tide correction is shown in the left figure and the number of valid Cryosat-2 observations with valid ocean tide correction using the FES2014 ocean tide model is illustrated in the right figure. Global testing showed that FES2014 seems to provide similar accuracy to the GOT4.8 ocean tide model and hence, we updated the ocean tide correction to FES2014 for DTU17 to retrain more data close to the coast.

To complete the DTU17 global marine gravity then north of 88°N and on land the DTU17 marine gravity field was augmented with EGM2008 free-air gravity ensuring global coverage. This somewhat older global marine gravity field was chosen to be consistent with the geoid used for the remove/restore process in deriving the gravity field.

2 Evaluation with Marine Gravity Observations

The standard evaluation with more than 1.4 million high quality edited un-classified marine gravity observations from the National Geospatial-intelligence Agency (NGA) was used to evaluate the various available global marine gravity fields in the northwest part of the Atlantic Ocean between 20°N and 45°N and 270°E and 330°E with statistics shown in Table 1. For comparison the Sandwell and Smith marine gravity field release 23.1 and 24.1 (Sandwell et al. 2013), available from http://topex.ucsd.edu/marine_grav/mar_grav.html, were also included.

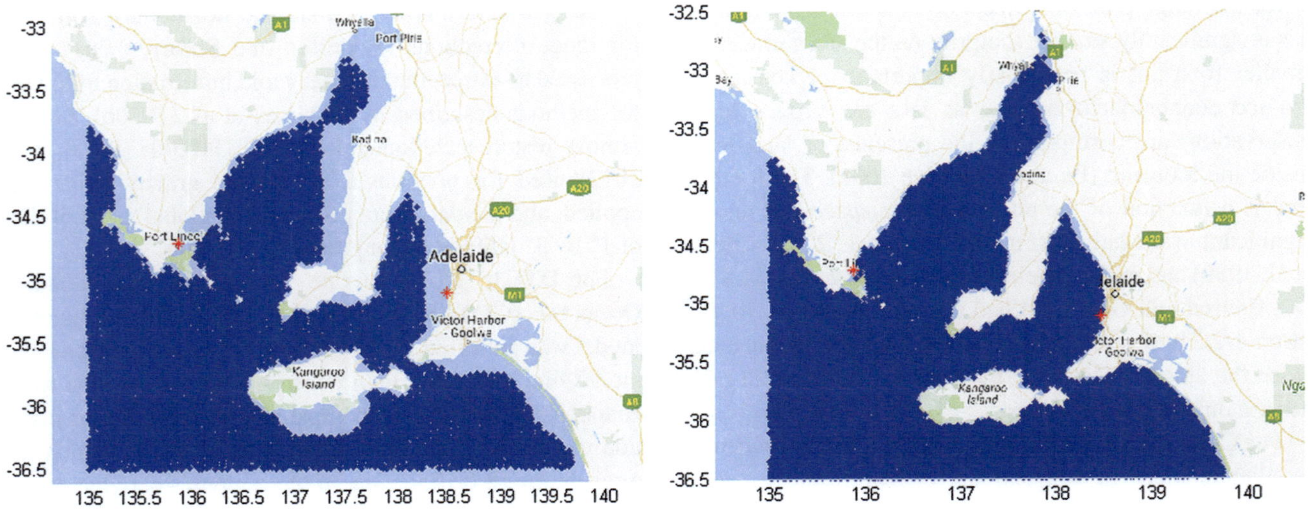

Fig. 1 The location of Cryosat-2 observations around Adelaide with the GOT4.8 ocean tide correction (left) and the FES2014 ocean tide model (right) applied

Table 1 Comparison with more than 1.4 million quality controlled marine gravity field observations in the northwest Atlantic Ocean

	Std. dev. (mGal)	Mean (mGal)	Max (mGal)
DTU17	2.51	0.5	32.4
DTU15	2.51	0.5	32.3
DTU13	2.83	0.5	32.2
DTU10	3.16	0.5	44.1
SS 23.1	3.13	0.7	43.4
SS 24.1	3.11	0.7	41.9

The comparison between altimetry and marine gravity also includes errors in the marine gravity observations. The accuracy of the marine gravity field observations is quoted by NGA to be around 1.5–2 mGal. With an observed standard deviation of 2.5 mGal between DTU15/DTU17 and the marine gravity field observations, this means that the accuracy of the altimetric gravity field must be around 2 mGal in this region.

The inclusion of additional data and application of slightly less spatial filtering does not improve the global statistics with marine gravity field observations for DTU17 vs the older DTU15. To investigate the DTU17 improvement a bit further, we have split the comparison with the NGA marine data into sub-comparisons by depth. From these sub-comparisons shown in Table 2, it is evident, that the largest improvement is seen in coastal regions where the impact of the SARAL/AltiKa can be seen.

On the contrary, the slightly less filtering applied in DTU17 has the effect of degrading the gravity field estimation in the depth range of 500–2,000 m. This is generally where the Gulf Stream is found in the northwest Atlantic Ocean and where the sea surface height variability is the highest (Andersen et al. 2010a).

Table 2 Sub-comparison by depth with the NGA data used in Table 1

	DTU17	DTU15	DTU17-DTU15 improvement
Depth (0–25 m)	2.25	2.32	4
Depth (25–50 m)	2.35	2.38	2
Depth (50–100 m)	2.48	2.54	3
Depth (100–500 m)	2.70	2.77	3
Depth (500–1,000 m)	2.55	2.52	−2
Depth (1,000–2,000 m)	2.57	2.52	−3
Depth (2,000–5,000 m)	2.55	2.55	0

Each depth interval has more than 40,000 marine observations for the comparison. The table presents the standard deviation (in mGal) for DTU17 and DTU15 as well as the improvement in percent

3 Marine Gravity Comparison in the Arctic Ocean

Due to the presence of sea ice, the number of ship-surveys is very limited in the Arctic Ocean, so we have performed a comparison with an international airborne survey called LomGrav. The survey was flown and conducted in 2009 by the DTU airborne system (Olesen 2003) in the mostly densely ice-covered part of the Arctic Ocean between Greenland and the North Pole. A little more than 55,000 high quality airborne observations were recorded within the region north of Greenland bounded by 80°N–90°N latitude and by 240°E–360°E longitude. Due to favorable flight conditions, the internal error at crossing points was less than 2 mGal on average (Olesen, personal communication, 2018). In this following analysis, the Sandwell and Smith gravity fields could not be included as the spatial coverage is limited to 80°N for these fields.

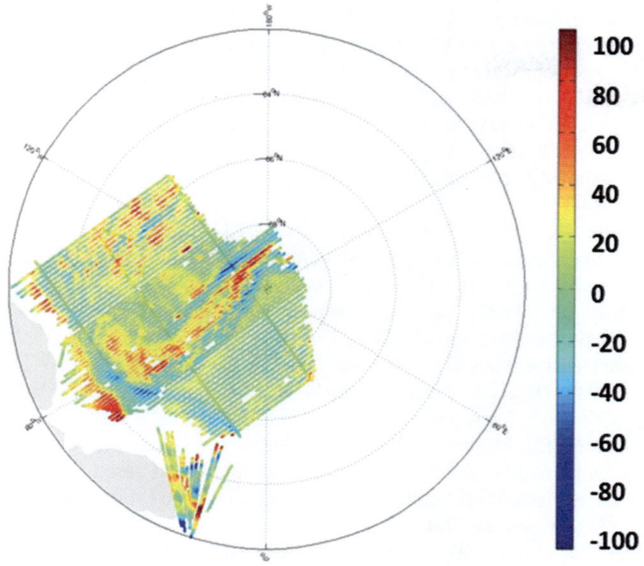

Fig. 2 The LomGrav2009 airborne marine gravity survey north of Greenland in the Arctic Ocean. Free-air gravity ranges from −100 to 100 mGal. The northern tip of Greenland is seen in grey

Table 3 Comparison with the roughly 55,000 airborne gravity observations in the LomGrav 2009 survey north of Greenland in the Arctic Ocean

All in mGal	EGM 2008	DTU10	DTU13	DTU15	DTU17
DTU17	9.82	8.81	5.91	5.45	3.78

The standard deviation of the Free-air gravity differences are shown in mGal

The LomGrav 2009 airborne gravity data are shown in Fig. 2. We compared the altimetric gravity with the airborne marine gravity by means of spline interpolation in the altimetric grid to the location of the airborne observations. The standard deviation of the differences for the various gravity fields are presented in Table 3.

Improvement of more than 50% in terms of standard deviation from pre to post Cryosat-2 launch in 2010 is evident. For the pre Cryosat-2 marine gravity fields, we found standard deviations of differences of 9.8 and 8.8 mGal for EGM2008 and DTU10, respectively. For the gravity field including the Cryosat-2, we found values of 5.9 mGal for DTU13 (1 year of C2 data), 5.45 mGal for DTU15 (4 years of C2 data), and 3.8 mGal for DTU17 (6 years of C2 data).

Even though the improvement is substantial, the comparison is somewhat larger than the 2 mGal quoted in the comparison with the NGA marine gravity data in the Atlantic and the 2 mGal internal consistency on the airborne gravity observations. This is very much expected, as the Arctic Ocean north of Greenland is among the most heavily sea ice covered regions of the world and we are only able to determine the sea surface height in sparse leads within the ice, when ice-movements cause it to crack open for a while. The nearly permanent sea ice covered regions are among the toughest regions for gravity field recovery from satellite altimetry.

4 Conclusion

With three geodetic missions, all providing data with higher range precision than the older ERS-1 and Geosat geodetic mission, altimetric gravity field accuracy is still increasing. The mapping of spatial wavelength within the 10–15 km range has increased dramatically revealing both new gravity field structures (Stenseng and Andersen 2012) but equally important revealed related bathymetric signals (Sandwell et al. 2014). Even though DTU15 and DTU17 have similar standard deviation with marine gravity in the northwest Atlantic Ocean, DTU17 offers improvement in the coastal zone due to the use of an improved ocean tide model. DTU17 also offers significant improvement in the Arctic Ocean due to longer time-series and improved data processing.

In the Arctic Ocean, this significant development in accuracy of the altimetric marine gravity is shown through comparison with high quality airborne data flown north of Greenland in 2009. An improvement of more than 50% in terms of standard deviation with the airborne data was seen in comparison with older gravity fields like DTU10 and EGM08, which are the only available global marine gravity field available within the region between Greenland and the North Pole.

Several interesting developments from these new data are still to come in the near future. In July 2017, the Jason-2 satellite initiated a 3-year Extension-of-Life (EoL) geodetic mission. The EoL will initially consist of two interleaved geodetic orbits of around 400 days repeat. This should decrease the cross-track difference for geodetic missions below 8 km for the first time, and bring the cross-track difference all the way down to 4 km in 2019. This will enable modelling of shorter wavelength in the gravity field to be mapped. If the Jason EoL last beyond this, another 2 interleaved geodetic orbits can bring the cross-track density down to 2 km by year 2021.

5 Data Availability

The DTU17 Global 1 min marine gravity field along with the DTU suite of related geophysical products like bathymetry and mean sea surface is available for research purposes from ftp.space.dtu.dk/pub/DTU17/ or by email request to the author at oa@space.dtu.dk.

References

Andersen OB, Scharroo R (2011) Range and geophysical corrections in coastal regions: and implications for mean sea surface determination. In: Vignudelli S et al (eds) Coastal altimetry. Springer, New York, pp 103–145. https://doi.org/10.1007/978-3-642-12796-0_5

Andersen OB, Knudsen P, Berry P (2010a) Recent development in high resolution global altimetric gravity field modeling. Lead Edge 29(5):540–545. ISSN: 1070-485X

Andersen OB, Knudsen P, Berry PAM (2010b) The DNSC08GRA global marine gravity field from double retracked satellite altimetry. J Geod 84:191–199. https://doi.org/10.1007/s00190-009-0355-9

Andersen OB, Knudsen P, Kenyon S, Factor JK, Holmes S (2017) Global gravity field from recent satellites (DTU15) – Arctic improvements. First Break 35(12):37–40. https://doi.org/10.3997/1365-2397.2017022. ISSN: 0263-5046

Carrere L, Lyard F, Cancet M, Guillot A, Picot N, Dupuy S (2015) FES2014: a new global tidal model. Presented at the Ocean Surface Topography Science Team meeting, Reston. Description at https://datastore.cls.fr/catalogues/fes2014-tide-model/

Chelton DB, Walsh EJ, MacArthur JL (1989) Pulse compression and sea level tracking in satellite altimetry. J Atmos Ocean Technol 6(1989):407–438. https://doi.org/10.1175/1520-0426(1989)006<0407:PCASLT>2.0.CO;2

Dibarboure G, Boy F, Desjonqueres JD, Labroue S, Lasne Y, Picot N, Poisson JC, Thibaut P (2014) Investigating short-wavelength correlated errors on low-resolution mode altimetry. AMS. https://doi.org/10.1175/JTECH-D-13-00081.1

Fu L-L, Cazenave A (eds) (2001) Satellite altimetry and earth sciences: a handbook of techniques and applications, Int. Geophys. Ser., vol 69. Academic, San Diego. 469 pp, ISBN: 9780080516585

Jain M, Andersen OB, Dall J, Stenseng L (2015) Sea surface height determination in the Arctic using Cryosat-2 SAR data from primary peak empirical retrackers. Adv Space Res 55(1):40–50. ISSN: 0273-1177. https://doi.org/10.1016/j.asr.2014.09.006

Olesen AV (2003) Improved airborne scalar gravimetry for regional gravity field mapping and geoid determination. Technical report, 24. National Survey and Cadastre, Copenhagen, 54 pp. ISBN: 87-7866-383-0

Pavlis NK, Holmes S, Kenyon S, Factor JK (2012) The development and evaluation of the Earth Gravitational Model 2008 (EGM2008). J Geophys Res. https://doi.org/10.1029/2011JB008916

Raney RK (1998) The delay Doppler radar altimeter. IEEE Trans Geosci Remote Sens 36:1578–1588

Sandwell DT, Garcia E, Soofi K, Wessel P, Chandler M, Smith WHF (2013) Towards 1-mGal accuracy in global marine gravity from Cryosat-2, Envisat and Jason-1. Lead Edge 32(8):892–899

Sandwell DT, Müller RD, Smith WH, Garcia E, Francis R (2014) New global marine gravity model from CryoSat-2 and Jason-1 reveals buried tectonic structure. Science 346(6205):65–67. https://doi.org/10.1126/science.1258213

Stammer D, Ray RD, Andersen OB, Arbic BK, Bosch W, Carrère L, Cheng Y, Chinn DS, Dushaw BD, Egbert GD, Erofeeva SY, Fok HS, Green JAM, Griffiths S, King MA, Lapin V, Lemoine FG, Luthcke SB, Lyard F, Morison J, Müller M, Padman L, Richman JG, Shriver JF, Shum CK, Taguchi E, Yi Y (2014) Accuracy assessment of global barotropic ocean tide models. Rev Geophys 52(3):243–282. https://doi.org/10.1002/2014rg000450. ISSN: 8755-1209

Stenseng L, Andersen OB (2012) Preliminary gravity recovery from CryoSat-2 data in the Baffin Bay. Adv Space Res 50(8):1158–1163

Global and Regional Evaluation of the First Year of Sentinel-3

Possibilities and Challenges for MSS Determination

Heidi Ranndal, Ole B. Andersen, and Per Knudsen

Abstract

The new Synthetic Aperture Radar (SAR) data from the Sentinel-3 satellites will provide the community with valuable new information in coastal areas and in the Arctic, due to the higher along-track resolution obtained through the Delay-Doppler processing. The SAR data also allows for a more detailed study of the ocean surface, since these make small-scale variations visible. Combined, data from the Sentinel-3 satellites creates a tremendous possibility for improving tidal models and mean sea surfaces near the coast, where these models are currently using extrapolation to provide information. However, some challenges are also becoming more apparent in areas where satellite altimetry have not previously been available. Such as the discrepancies between tidal models near the coast, which are amplified because the Sentinel-3 satellites fly in a sun-synchronous orbit. Acquiring satellite altimetry data in coastal and sea ice covered areas also highlights some issues with the current wet tropospheric correction, calculated from measurements by the on-board microwave radiometer, leading us to the conclusion that it is safest to use a WTC from a model – at least in coastal and sea ice prone areas.

Keywords

Coastal altimetry · FES2014 · GOT4.10 · MSS · Sentinel-3 · SLA

1 Introduction

Sentinel-3A was launched in 2016 and is the first Synthetic Aperture Radar (SAR) only altimetric mission. It has a repeat period of 27 days giving us data along new ground tracks, which is very valuable for Mean Sea Surface (MSS) evaluation. The orbital setup and instrumental capabilities of Sentinel-3 provide the altimetric community with a new set of opportunities and challenges. Some of these are highlighted in this paper, such as the amplification of tide model discrepancies due to the sun-synchronous orbit of the satellite, and small-scale variations in the sea level anomalies that are visible due to the SAR altimeter. SAR altimetry has a high along-track resolution, which provides height estimates much closer to the coast, although the altimeter waveforms, due to the shape of the altimeter footprint, might be more susceptible to contamination of various types, such as swell (Moreau et al. 2018) and land reflections depending on the orientation of the ground track with respect to the coastline (Dinardo et al. 2011). In general, ground tracks that lie parallel to the coastline will be more prone to land contamination.

2 Data and Methods

The data used for this study were downloaded from the Radar Altimetry Database System (RADS) (Scharroo et al. 2016) and consist mainly of cycles 8–21 of Sentinel-3A in 1 Hz, which corresponds to 1 year of data. For comparisons, data from Satellite with ARgos and ALtiKa (SARAL)

H. Ranndal (✉) · O. B. Andersen · P. Knudsen
Danmarks Tekniske Universitet, Lyngby, Denmark
e-mail: hvil@space.dtu.dk; oa@space.dtu.dk

S. P. Mertikas, R. Pail (eds.), *Fiducial Reference Measurements for Altimetry*,
International Association of Geodesy Symposia 150, https://doi.org/10.1007/1345_2018_50

and Envisat have also been downloaded from RADS. For Envisat, cycles 77–89 were used, and for SARAL, cycles 33–46 were used. Unless otherwise stated, the Sea Level Anomaly (SLA) is referenced to the DTU15MSS (Andersen et al. 2016). By using RADS data we ensure comparability between the data sets, since biases have been minimized and geophysical corrections are consistent. However, RADS only provides 1 Hz data, so the benefits of using the 20 Hz data from the Sentinel-3A SAR altimeter will not be studied here.

3 Results

3.1 Mapping Signals with Shorter Wavelengths

With Sentinel-3A SAR altimetry the community is suddenly able to study the global ocean in a much higher resolution. The short wavelength noise and the ability of the SAR altimeter to map short wavelength signals is investigated from 1 Hz data from Sentinel-3A. A couple of comparisons are shown in Figs. 1 and 2 for pass #51 of Sentinel-3A and pass #681 for Envisat and SARAL/AltiKa. 14 cycles are shown for each mission. The short wavelength SLA has been derived by subtracting a moving average of 15 along-track (the current location and the adjacent 14) measurements, in order to filter out signals with wavelengths longer than roughly 100 km. Sentinel-3A shows a clear signal in the highlighted area, which is not visible in SARAL or Envisat data. 20 Hz data were also briefly compared, but exhibited too much noise to identify the same signal without further processing or introduction of other methods.

3.2 Tide Models and MSS from Sentinel-3

Due to the higher along-track resolution of the SAR altimeter, it is possible to retrieve a lot more data closer to the coasts compared to conventional altimetry missions. Figures 3 and 4 show the difference between FES2014 (Carrère et al. 2016) and GOT4.10 (Ray 2013) for the 14 cycles of Sentinel-3A data. In the open ocean, where there has already been taken advantage of the conventional altimetry, the two models agree well. However, the difference between the two tidal models is several centimetres in coastal areas and in regions prone to sea ice cover. With the use of the new SAR data from Sentinel-3A, the models can be significantly improved in these areas. Especially if the information from the 20 Hz data is utilized.

The sun-synchronous orbit of Sentinel-3A amplifies the errors in the S2 constituent of the tidal models, since the satellite always sees the same location at the same time of day. The SLA field from Sentinel-3A data will therefore, in an even higher degree than data from other missions, depend on the chosen tidal model. Also, the influence of tide models on MSS determination should be considered. As S2 residuals might map directly into the MSS when deriving a MSS from Sentinel-3 data, we tried to quantify the possible influence of the erroneous S2 residual tide signal that could go into the MSS. This was done by studying the difference of using ocean tide corrections from two state of the art ocean tide models. Due to the orbit of Sentinel-3, this difference will correspond to an upper bound of the effect of the erroneous S2 residual. The standard deviation of the difference between the GOT4.10 and the FES2014 ocean tide corrections is 2.7 cm for the data shown in Fig. 3. The effect will be much smaller at lower latitudes where satellite altimetry is also available from the Jason

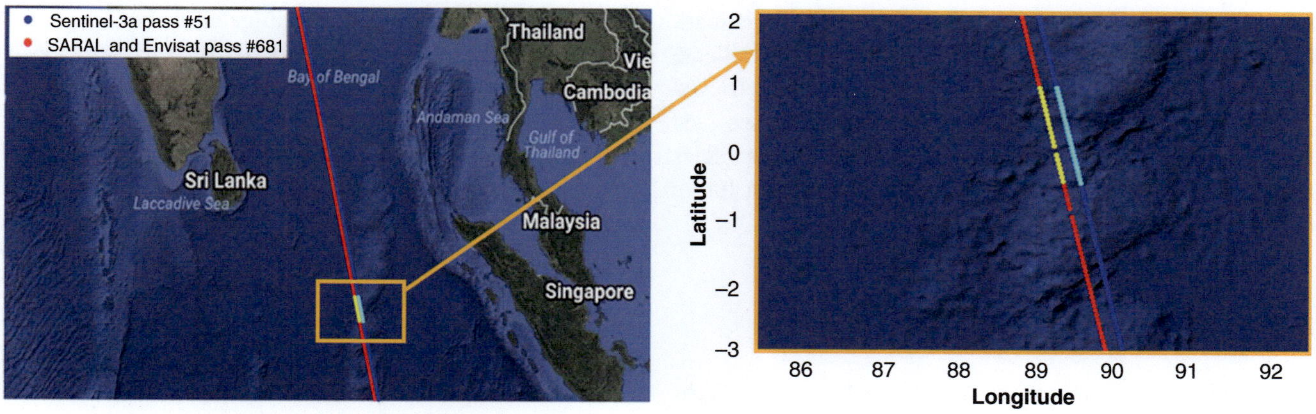

Fig. 1 Location of the passes shown in Fig. 2. The highlighted part of the passes (cyan and yellow) corresponds to the area plotted in Fig. 2

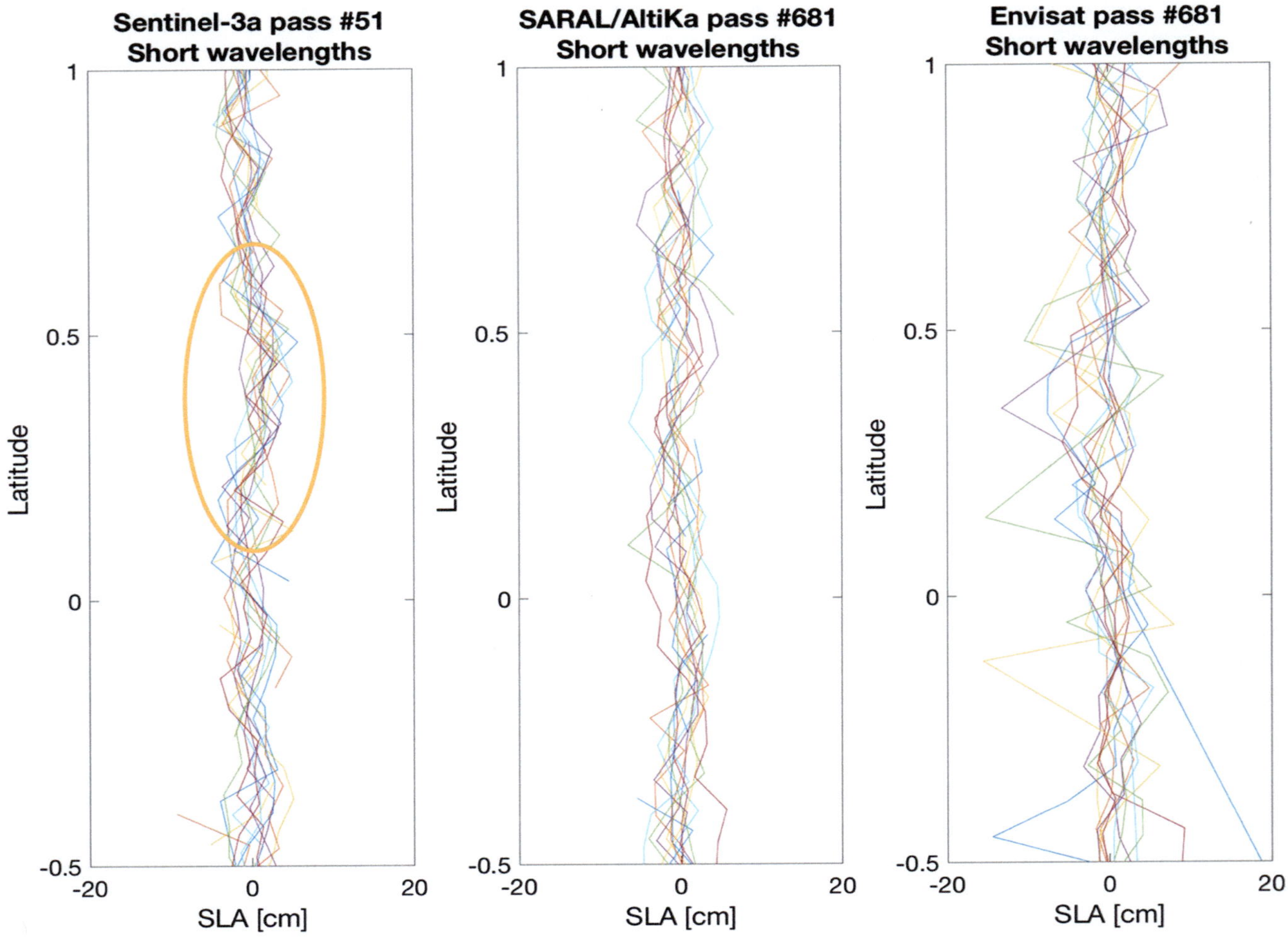

Fig. 2 Comparison of altimetry data obtained by three different satellites from adjacent passes in the Indian Ocean. Sentinel-3A SAR data (left), SARAL ka-band data (middle) and Envisat data (right)

satellites, which are not sun-synchronous. However, in the Arctic ocean, only sun-synchronous satellites are available so here it is likely more representative of the potential error.

Besides the tidal models, differences are also found in the mean sea surfaces from the Technical University of Denmark (DTU) and Centre National D'Études Spatiales - Collecte Localisation Satellites (CNES-CLS), when data are retrieved close to the coast. Two examples of affected passes (see locations in Fig. 5) are shown in Fig. 6. These discrepancies can be minimized if new SAR data are included in future derivations of the MSS.

3.3 Wet Tropospheric Correction

An unprecedented amount of near-coastal data has highlighted an instrumental issue with the Microwave Radiometer (MWR) carried on board the Sentinel-3A satellite, which estimates the Wet Tropospheric Correction (WTC). Preliminary studies of the Sentinel-3A data showed a general trend towards high SLA values close to the coast. This was found to be caused by an underestimation of the WTC from the on-board radiometer, since the radiometer carried by Sentinel-3A is sensitive to land contamination near the coast as it has a larger footprint than the SAR altimeter in the along-track direction (Andersen and Scharroo 2010). It has previously been shown that land contamination of the MWR happens as far as 20–25 km from the coast (Fernandes and Lázaro 2018). This effect can be seen in Fig. 7, where the mean difference between the measured and a modelled WTC is shown. For this comparison, we chose the correction from the European Centre for Medium-Range Weather Forecasts (ECMWF) atmospheric model. Near the coast, the underestimation of the MWR WTC can be up to 20 cm, causing incorrectly high SLA values. In the open ocean, the correction from the radiometer is usually slighter higher than the one from ECMWF. The issue is also present in sea ice covered areas, which can be seen in Fig. 8, where the WTC difference can be up to 50 cm. For both coastal areas and sea ice

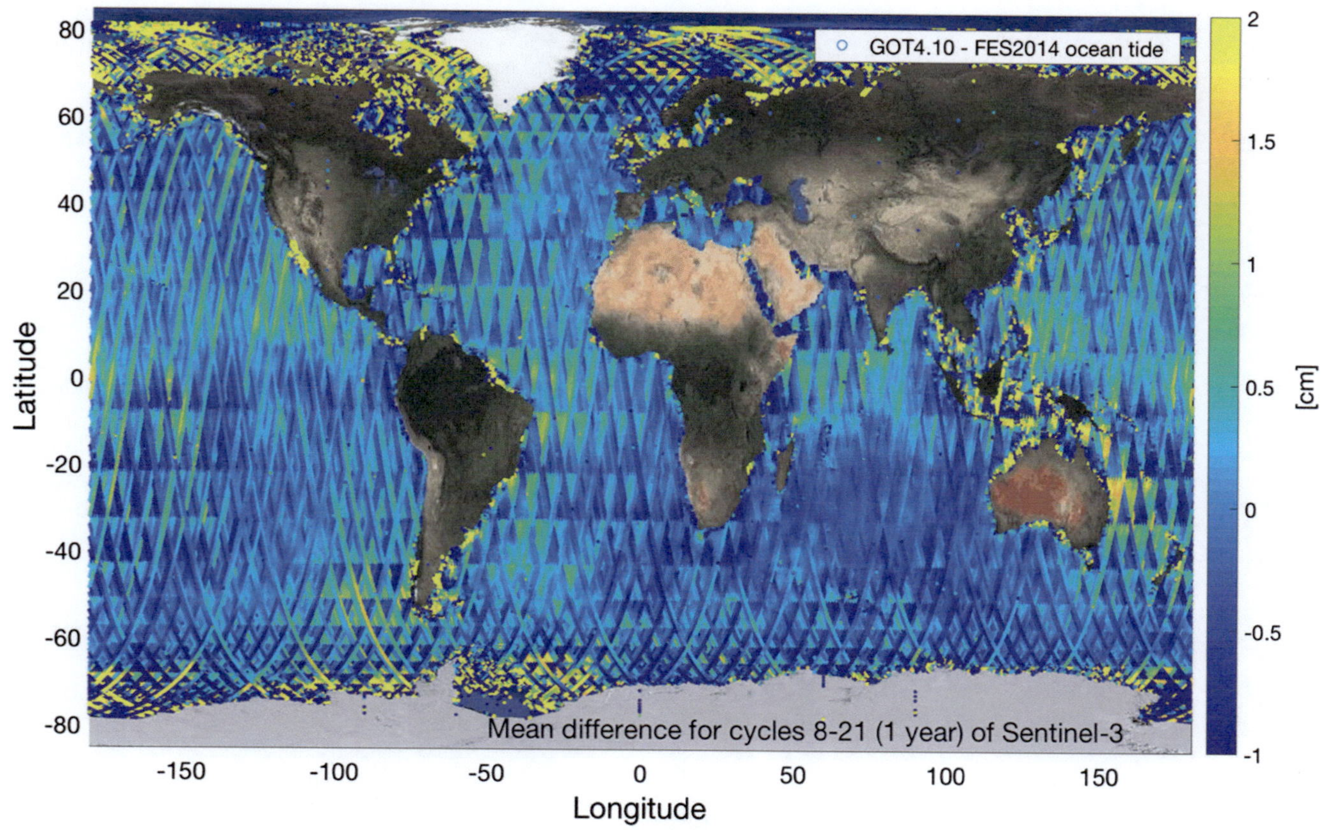

Fig. 3 Global difference between 1 year mean of ocean tide corrections when using tide models FES2014 and GOT4.10. Sentinel-3A data for cycles 8–21 were obtained through RADS

covered regions, the estimated correction from the MWR becomes negative and is therefore not big enough, causing too high SLA estimates in these regions. For SAR data, it was therefore found necessary to use the wet tropospheric correction derived by the ECMWF atmospheric model. As an alternative, a method such as described in Fernandes et al. (2010, 2015) could be used, so that the wet tropospheric correction could be obtained through a Global Navigation Satellite System (GNSS) derived Path Delay (GPD) algorithm, where both coastal GNSS stations, atmospheric models, and MWR measurements are used to achieve an improved estimate of the WTC. This would alleviate many of the erroneous WTC estimates from the radiometer along the coast. A thorough analysis of the WTC for Sentinel-3A measurements has been performed in Fernandes and Lázaro (2018).

4 Conclusions

Due to the high along-track resolution of the Sentinel-3 satellites, they have been expected to provide new and valuable information about the surface topography in a much higher resolution than possible before. Not only will the satellites measure the height of the sea surface, waves and wind speed over the oceans, but they will also provide measurements over sea ice, ice sheets, rivers, and land with an unprecedented accuracy. From the results shown in this study, it is clear that a new era of global SAR altimetry will provide valuable information about small-scale variations of the sea surface and about the sea level changes close to the coast, which will assist the improvement of the MSS especially in coastal areas, where extrapolation will no longer be necessary. Improving the MSS near the coast is crucial for a precise calibration between satellite altimetry and in-situ data from fiducial reference measurements. However, it is important to determine how many of our models and corrections are of a satisfactory quality near the coast and in Arctic regions, where conventional altimetry data has not been available, and where errors might not yet have been found and corrected. As shown here and in previous studies, one major error source is the MWR WTC, which can cause too high SLA estimates near the coast and at high latitudes. Another issue is the sun-synchronous orbit, which will amplify any errors and discrepancies of the S2 component in the tidal models. These challenges need to be quantified and addressed before using Sentinel-3 data

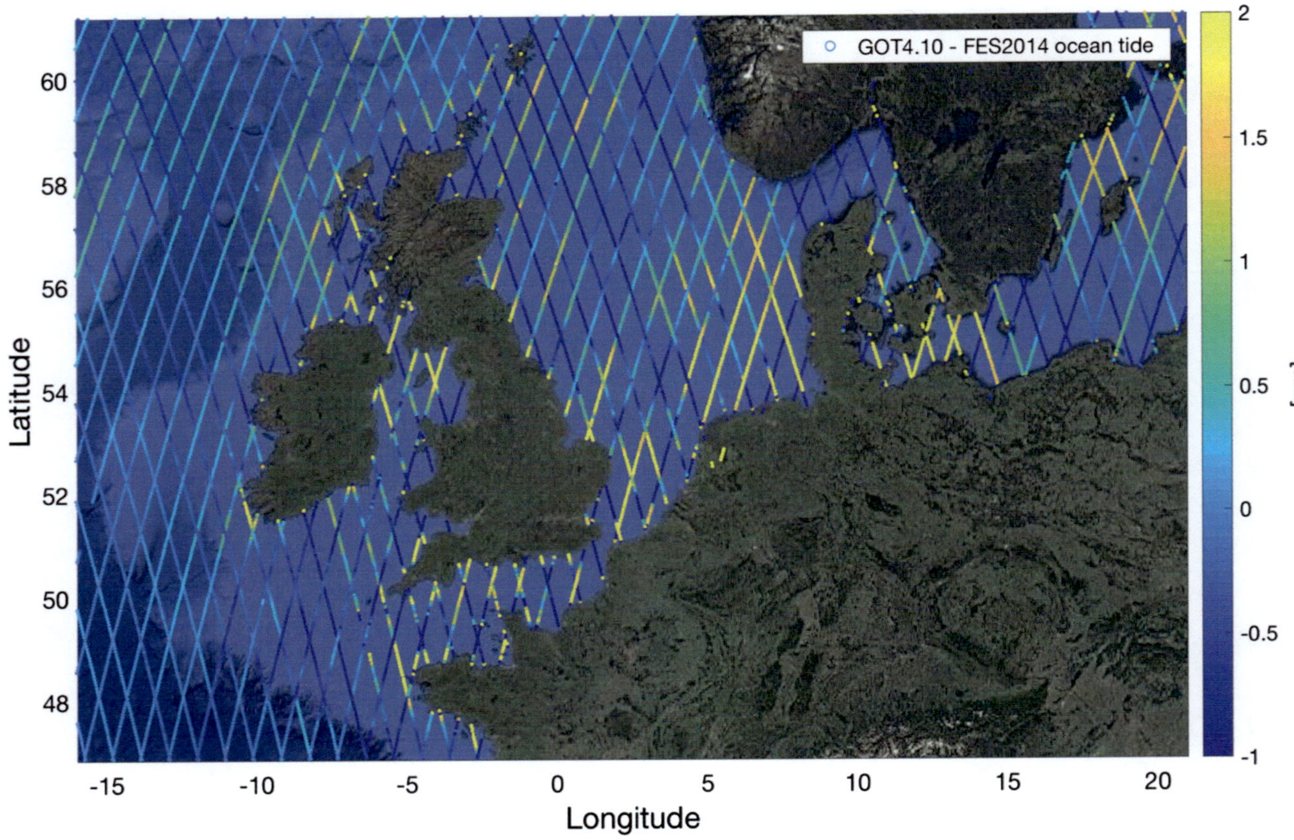

Fig. 4 Difference between 1 year mean of ocean tide corrections when using tide models FES2014 and GOT4.10. Sentinel-3A data for cycles 8–21 were obtained through RADS

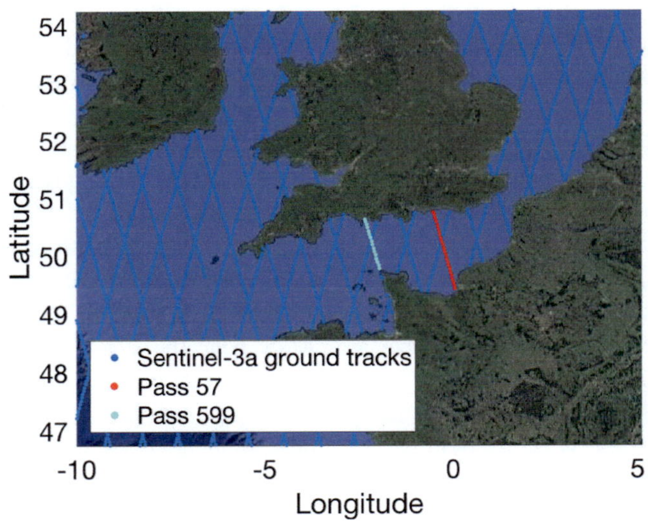

Fig. 5 Location of passes shown in Fig. 6

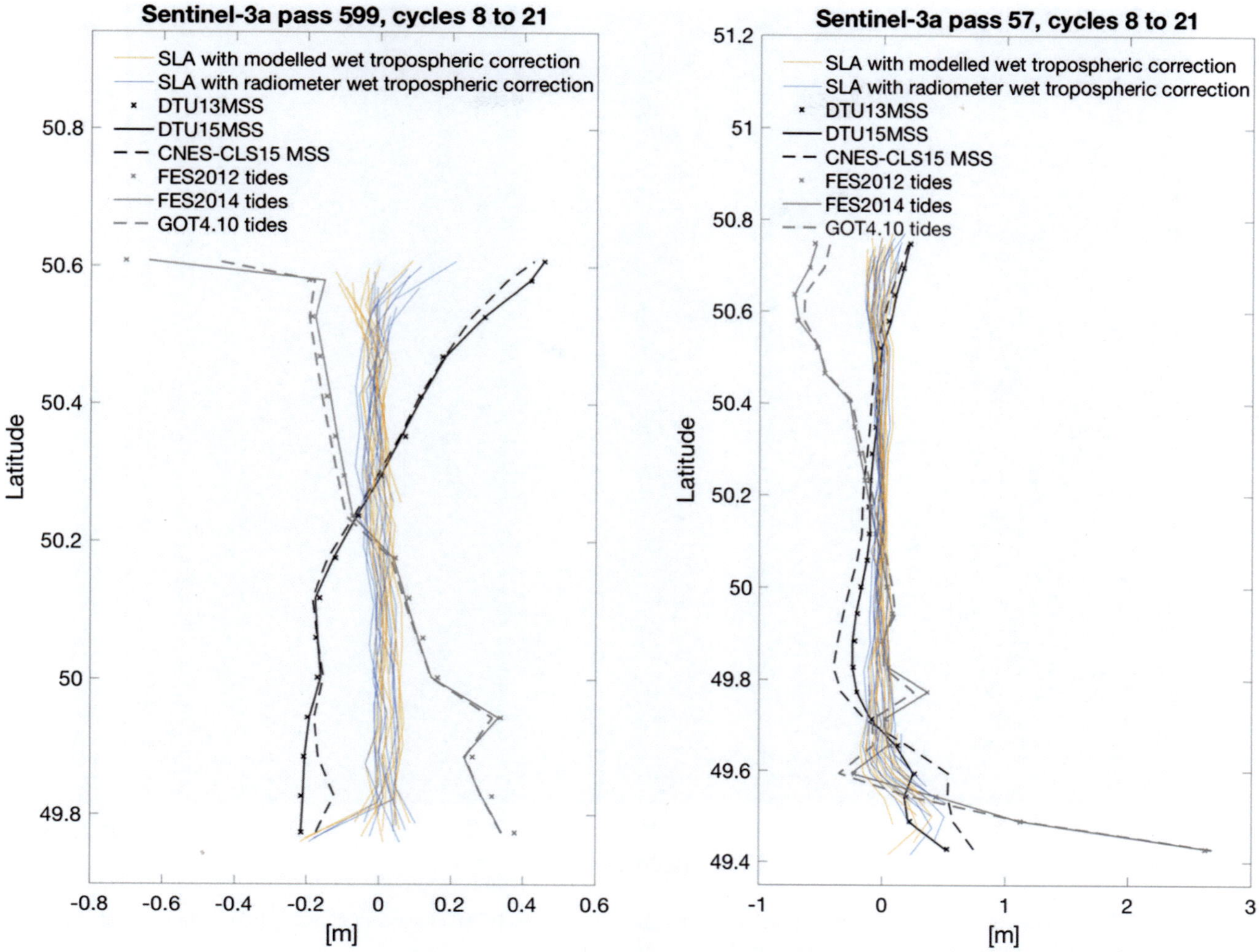

Fig. 6 Examples of different mean sea surfaces (black) and tidal models (grey) for two passes in the English Channel, along with SLA with either WTC from radiometer or the ECMWF model (SLA are only shown using FES2014 and DTU15MSS)

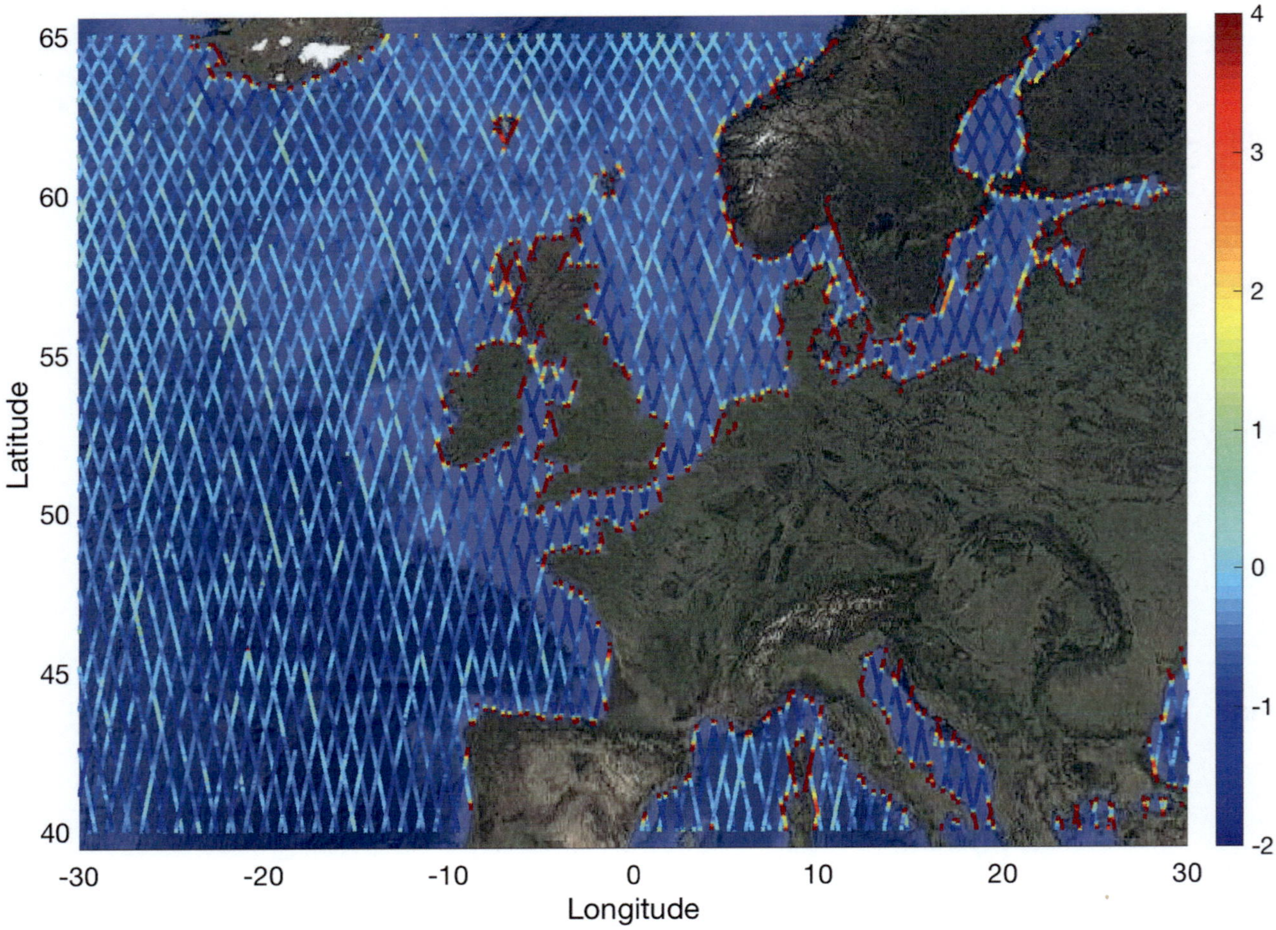

Fig. 7 Mean difference between wet tropospheric corrections from radiometer and the ECMWF model for cycles 8–21 of Sentinel-3A data in the North Sea. Positive values correspond to a higher correction from the ECMWF model

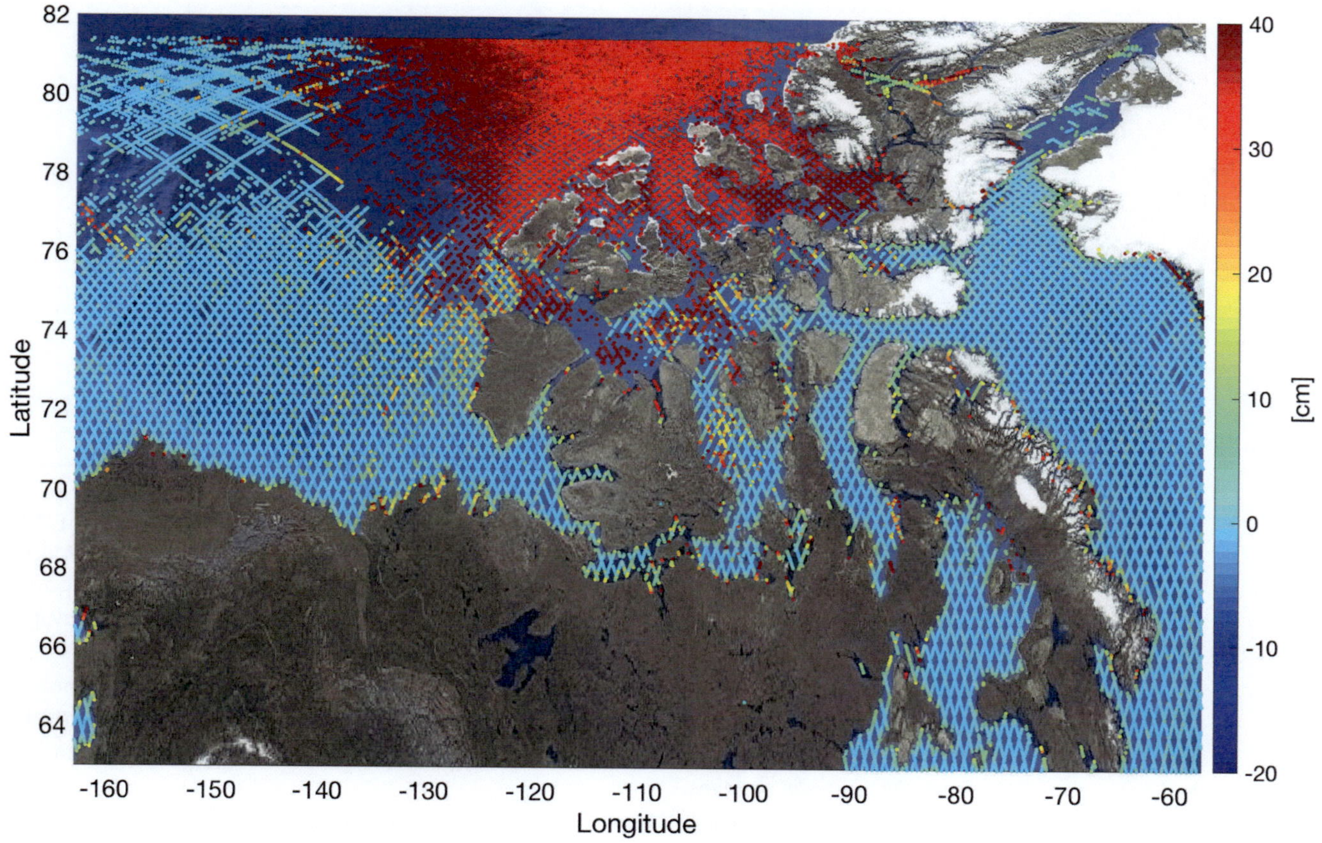

Fig. 8 Mean difference between wet tropospheric corrections from radiometer and the ECMWF model for cycles 8–21 of Sentinel-3A data in the Canadian Arctic Ocean. Positive values correspond to a higher correction from the ECMWF model

for MSS determination or to draw any conclusions on the changing sea level.

Acknowledgements We would like to thank the European Space Agency for supporting the Fiducial Reference Measurements for Altimetry (FRM4ALT) project, which aims at establishing a continuous and reliable monitoring of sea level changes, by providing long-term fiducial reference measurements on the ground. We also thank the two anonymous reviewers for their valuable notes and comments.

References

Andersen OB, Scharroo R (2010) Range and geophysical corrections in coastal regions. In: Vignudelli et al (ed), Coastal altimetry. ISBN: 978-3-642-12795-3

Andersen OB, Stenseng L, Piccioni G, Knudsen P (2016) The DTU15 MSS (mean sea surface) and DTU15LAT (lowest astronomical tide) reference surface. In: ESA living planet symposium, Prague, Czech Republic. Available from http://ftp.space.dtu.dk/pub/DTU15/DOCUMENTS/MSS/DTU15MSS+LAT.pdf

Carrère L, Lyard F, Cancet M, Guillot A, Picot N (2016) FES 2014, a new tidal model—validation results and perspectives for improvements. In: Proceedings of the ESA living planet symposium 2016, Prague, Czech Republic, 9–13 May

Dinardo S, Lucas B, Benveniste J (2011) SAR altimetry in coastal zone: performances, limits, perspectives. In: Fifth coastal altimetry workshop, presentation, San Diego. http://www.coastalt.eu/sandiegoworkshop11

Fernandes MJ, Lázaro C (2018) Independent assessment of sentinel-3A wet tropospheric correction over the open and coastal ocean. Remote Sens 10(3):484. https://doi.org/10.3390/rs10030484, http://www.mdpi.com/2072-4292/10/3/484.

Fernandes MJ, Lázaro C, Nunes AL, Pires N, Bastos L, Mendes VB (2010) GNSS-derived path delay: an approach to compute the wet tropospheric correction for coastal altimetry. Geosci Remote Sens Lett 7(3):596-600. https://doi.org/10.1109/LGRS.2010.2042425

Fernandes MJ, Lázaro C, Ablain M, Pires N (2015) Improved wet path delays for all ESA and reference altimetric missions. Rem Sens Env 169:50–74. https://doi.org/10.1016/j.rse.2015.07.023

Moreau T, Tran N, Aublanc J, Tison C, Le Gac S, Boy F (2018) Impact of long ocean waves on wave height retrieval from SAR altimetry data. Adv Space Res **62**(6), 1434–1444 https://doi.org/10.1016/j.asr.2018.06.004

Ray RD (2013) Precise comparisons of bottom-pressure and altimetric ocean tides. J Geophys Res 118:4570–4584. https://doi.org/10.1002/jgrc.20336

Scharroo R, Leuliette EW, Lillibridge JL, Byrne D, Naeije MC, Mitchum GT (2016) RADS: consistent multi-mission products. In: Proceedings of the symposium on 20 years of progress in radar altimetry, Venice. European space agency special publication, ESA SP-710, 4

Arctic Freshwater Fluxes from Earth Observation Data

Ole B. Andersen, Karina Nilsen, Louise S. Sørensen, Henriette Skourup, Natalia H. Andersen, Thomas Nagler, Jan Wuite, Alexei Kouraev, Elena Zakharova, and Diego Fernandez

Abstract

Through both atmospheric and oceanic circulation, heat is transferred between the equator and the poles. Possible ways in which the Arctic ecological systems can be affected by warmer temperatures include: changes in amount and duration of snow and ice cover; frequency and extent of spring floods; changes in the ratio of precipitation minus evapotranspiration; amounts of water transport from lakes and rivers from snow and permafrost melting; and a decrease in frozen precipitation. A key component in transferring heat is through freshwater exchange in and out of the Arctic. Hence, accurate mapping freshwater fluxes and potentially its changes with time is vital to describe the heat transfer and its possible temporal changes.

Our results demonstrate how ESA's Earth Observation data together with in-situ measurements can be used to improve the mapping of the major Arctic Ocean freshwater fluxes. In this paper, we outline how four of the five major freshwater fluxes can be determined using present day Earth Observation data exclusively. These are: discharge from rivers; inflow through ice and melt run off; outflow of freshwater in sea ice; and in/outflow of freshwater through ocean currents. We subsequently present key finding and estimates of these four freshwater fluxes and compare our results with estimates based on in-situ data provided through previous studies.

Keywords

Altimetry · Arctic freshwater · Earth observation

1 Introduction

The Arctic Ocean is sensitive to freshwater fluxes (FWF), where the main inflow of freshwater comes from river discharge, ice and snow discharge, net precipitation, and inflow of low-salinity water through the Bering Strait. The inflow is counter-acted through evapotranspiration and outflow, where the major pathways are the Fram Strait connected to the North Atlantic Ocean, and the Davis Strait connected to the Canadian Arctic Archipelago (CAA) and subsequently the North Atlantic Ocean. Arctic Freshwater balance plays a fundamental role in the Arctic cryosphere and ocean and consequently in climate research. Therefor changes to the

O. B. Andersen (✉) · K. Nilsen · L. S. Sørensen · H. Skourup · N. H. Andersen
Technical University of Denmark, DTU Space, Lyngby, Denmark
e-mail: oa@space.dtu.dk

T. Nagler · J. Wuite
ENVEO, Innsbruck, Austria

A. Kouraev
LEGOS/CNRS, Toulouse, France

E. Zakharova
IWP RAS, Moscow, Russia

D. Fernandez
ESA ESRIN, Frascati, Italy

S. P. Mertikas, R. Pail (eds.), *Fiducial Reference Measurements for Altimetry*,
International Association of Geodesy Symposia 150, https://doi.org/10.1007/1345_2019_75

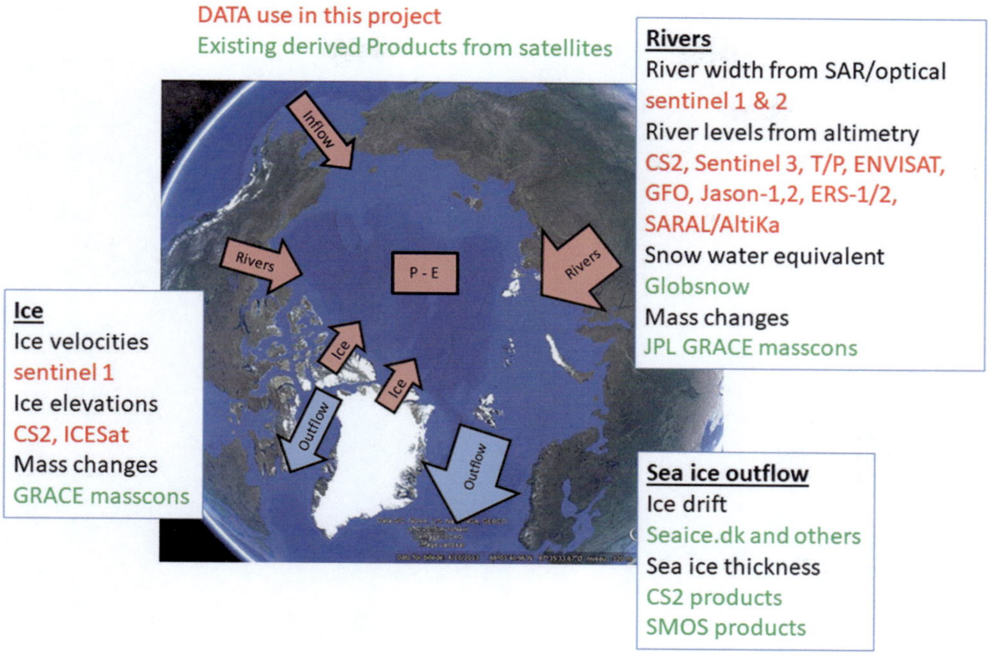

Fig. 1 Dominating freshwater fluxes transporting freshwater in and out of the Arctic Ocean. Boxes illustrate the major FWF and the Earth Observation data contributing to this study

balance of freshwater fluxes in and out of the Arctic Ocean might have large consequences (Oltmanns et al. 2018).

The main objective of this study has been to determine how the various components of FWF can be mapped with Earth Observation (EO) data together with in-situ data. The four FWF components, which have been studied, are illustrated with arrows in Fig. 1. These are: (1) Discharge from rivers, (2) Inflow from land-ice and melt run off, (3) outflow of sea ice, and (4) in- and outflow of freshwater in the ocean. Precipitation minus evapotranspiration (P-E) is also a key FWF and one of the largest components. However, P-E can only be studied on land using EO data but not over the ocean. Consequently, we determined it from models. Smaller FWF components are groundwater FWF and FWF from permafrost thawing. We omitted these in the current investigation and are currently an error source to our investigation. They will become important to study using EO data in the future.

Various authors and scientists have determined the FWF using models with only little involvement of EO data (Haine et al. 2015). A detailed summary of present day knowledge about the Arctic FWF and the Arctic Freshwater budget is outlined by Carmack et al. (2016) illustrating the magnitude and accuracy with which each freshwater components are known. As the summary by Carmack et al. 2016) is generally based on modelling studies based on in-situ data when available, it is a good baseline for evaluating the potential of EO data to map the associated FWF in the Arctic.

2 Land Ice and Melt FWF

The fresh water flux from the ice sheet and ice caps includes two major processes, (1) the calving flux from marine terminating glaciers and (2) melt water flux (run-off). The FWF from the land ice is the sum of discharge, and run-off. Discharge is determined from observations of ice velocity and estimates of ice thickness, while run-off is estimated from Regional Climate Model (RCM) output from the three regional climate models: (MAR (Gallee and Schayes 1994), HIRHAM (Bøssing Christensen et al. 2007) and RACMO (van Meijgaard et al. 2008).

To calculate the calving flux into the ocean, ice velocity and thickness are required where the glacier meets the ocean (the flux gate). We derived ice velocity from repeat pass SAR data (i.e. Sentinel-1) as illustrated in Fig. 2 (Nagler et al. 2015), while H is based on radar-derived ice thickness measurements using a mass conservation approach (Morlighem et al. 2017). As example, Fig. 3 shows annually averaged profiles of ice velocity and thickness along the flux gate of a drainage basin in the north of Greenland. This basin has an area of approximately 63,019 km^2 and includes Ryder, C. H. Ostenfeld and Steensby Glaciers as well as several smaller marine-terminating glaciers. Using a depth correction factor of 0.95 (Nagler et al. 2015) to derive depth-averaged velocity from the surface velocity maps, and assuming the flux through the selected gate equals the total discharge of the basin, the ice flux through the gate

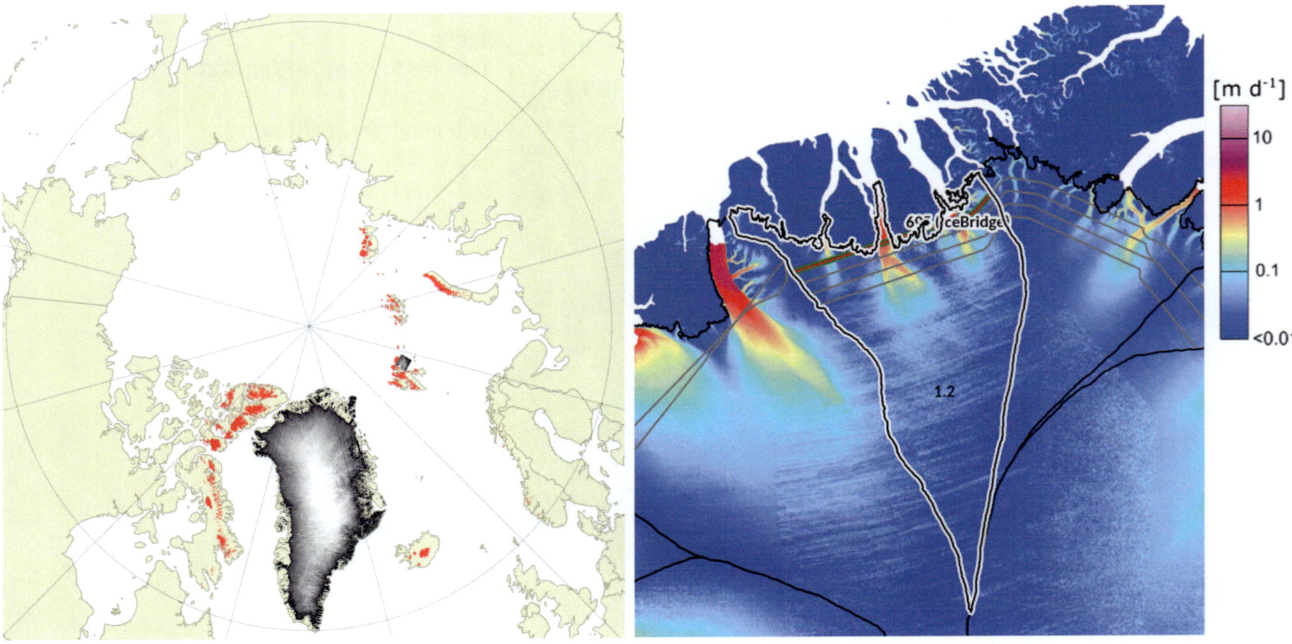

Fig. 2 (Left) Land ice coverage in the Arctic from RGI v.6 (RGI Consortium 2017) showing the ice thickness maps for Greenland and Austfonna (Svalbard). Areas with little or no or ice thickness data are depicted in red. (Right) Ice velocity map for north Greenland derived from Sentinel-1 SAR data (Nagler et al. 2015) showing major drainage basins (solid black lines) and Operation IceBridge RES flight lines (solid grey lines)

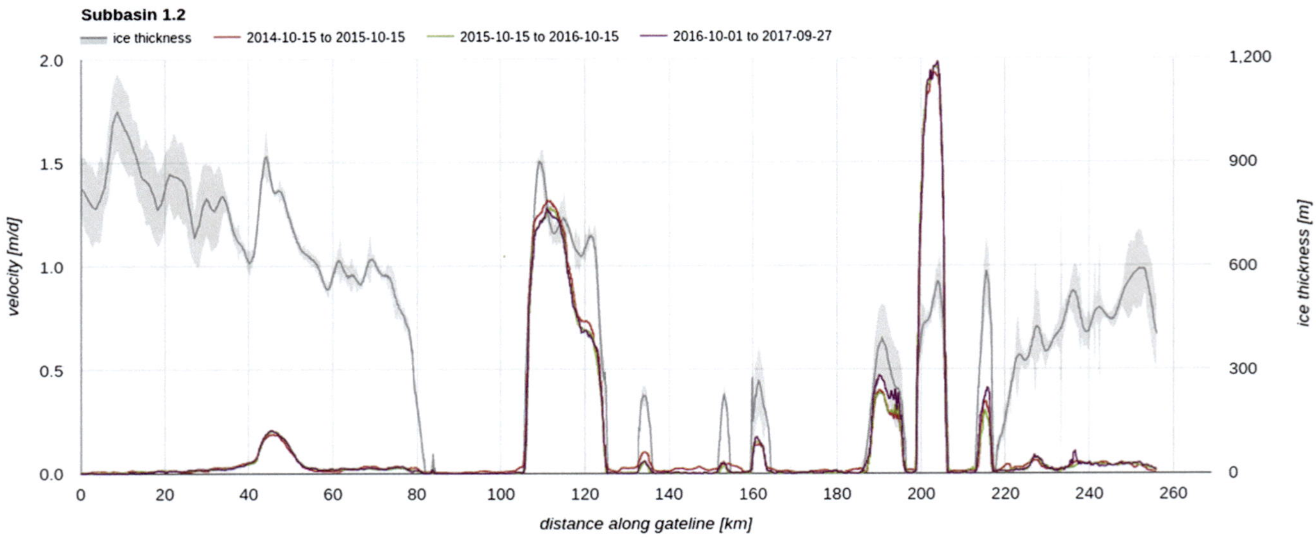

Fig. 3 Profiles of ice velocity and ice thickness (gray) along the gate of basin 1.2 (see red line in right panel of Fig. 2) and calculated ice flux and ice flux increase for three consecutive years (2014/15, 2015/16, 2016/17)

is calculated. The flux varies between 6.63 ± 0.68 km^3/year and 6.68 ± 0.70 km^3/year with less than 1% inter annual changes over the period 2014–2017. Run-off estimates are determined monthly, by taking the average of three available regional climate models MAR, HIRHAM and RACMO, and summed to provide yearly run-off estimates consistent with the solid ice discharge. For the example drainage basin, RU varies between 10.04 Gt/year and 10.30 Gt/year. Using an ice density of 917 kg/m^3, this yields a total annual FWF between 16.12 ± 1.39 Gt/year and 16.41 ± 1.32 Gt/year for the basin.

The applied method is considered state-of-the-art and has the capability to deliver unprecedented temporal sampling of ice flow velocity, and if ice thickness data is available, the method will also deliver freshwater flux in a consistent systematic and fully automated manner.

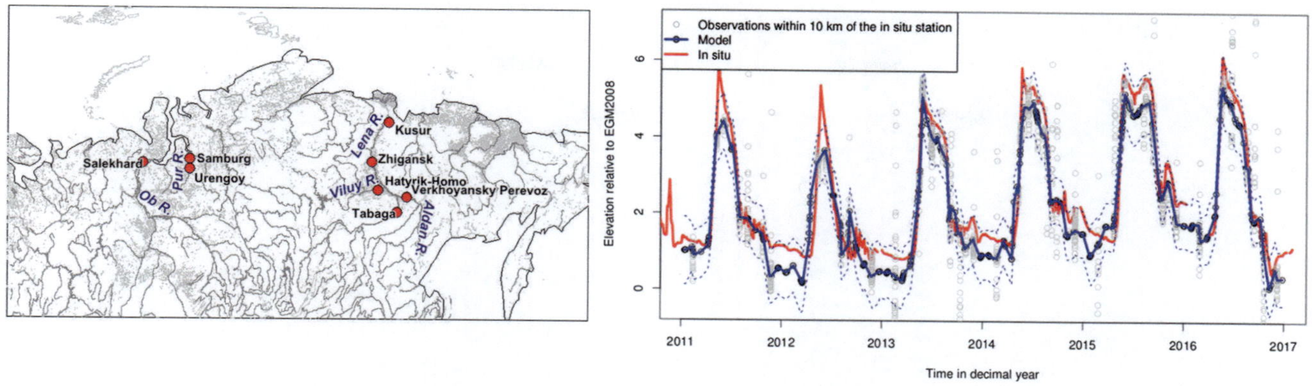

Fig. 4 (left) Location of the Ob river used for the estimation of river discharge. (right) River level of the Ob River in meters estimated using Cryosat-2 Synthetic Aperture Radar Altimetry (blue) along with the river gauge observations at Salekhard in red

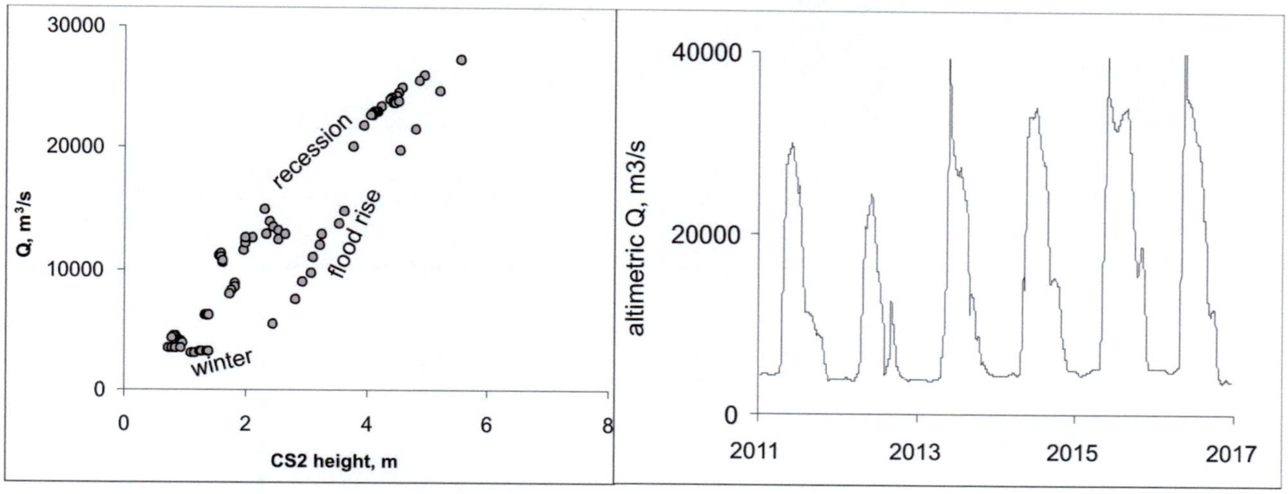

Fig. 5 (left) The relation between Cryosat-2 (CS2) height and in situ discharge for the Ob River. (right) Retrieved altimetric water discharge for the Ob River in m3/s

3 River FWF

River discharge are determined from the classical rating curve approach, where the water height, retrieved from altimetric satellites, is functionally related to the discharge on the gauge stations within the river (Kouraev et al. 2004). The river Ob was used because it has adequate width to produce a clear altimetric signal and because it is equipped with in-situ observations from the river gauge Selekhard. Figure 4 (left) shows the location and Fig. 4 (right) shows the modelled river height derived from Cryosat-2 and the river gauge at Selekhard. Blue dots represent all Cryosat-2 observations within 10 km and the solid blue line the interpolated modelled river height (Nielsen et al. 2015). The red color represent the Salekhard measured river heights.

From the in-situ observations of water level and discharge at the Selekhard river gauge, we could determine the rating curve, which determine the relationship between river height and discharge. In this way, we could determine the river discharge or freshwater flux using satellite altimetry with an accuracy of 2–7% for the Ob river illustrated in Fig. 5.

4 Sea Ice FWF

The sea ice volume flux is given by the sum of sea ice thickness and the depth of the snow layer across a pre-defined flux gate, together with the velocities of the sea ice drift perpendicular to the flux gate. The conversion of sea ice volume flux into equivalent freshwater content uses a priori knowledge of the densities of sea ice, snow, sea- and fresh water. The largest sea ice flux gate in the studied region is the Fram Strait (se location in Fig. 1).

With recent availability of sea ice thickness estimates from satellite radar altimeters and drift from repeated SAR imagery, we used a combination of these satellite-based observations to estimate the present FWF across the Fram Strait. Monthly sea ice volume fluxes and equivalent fresh

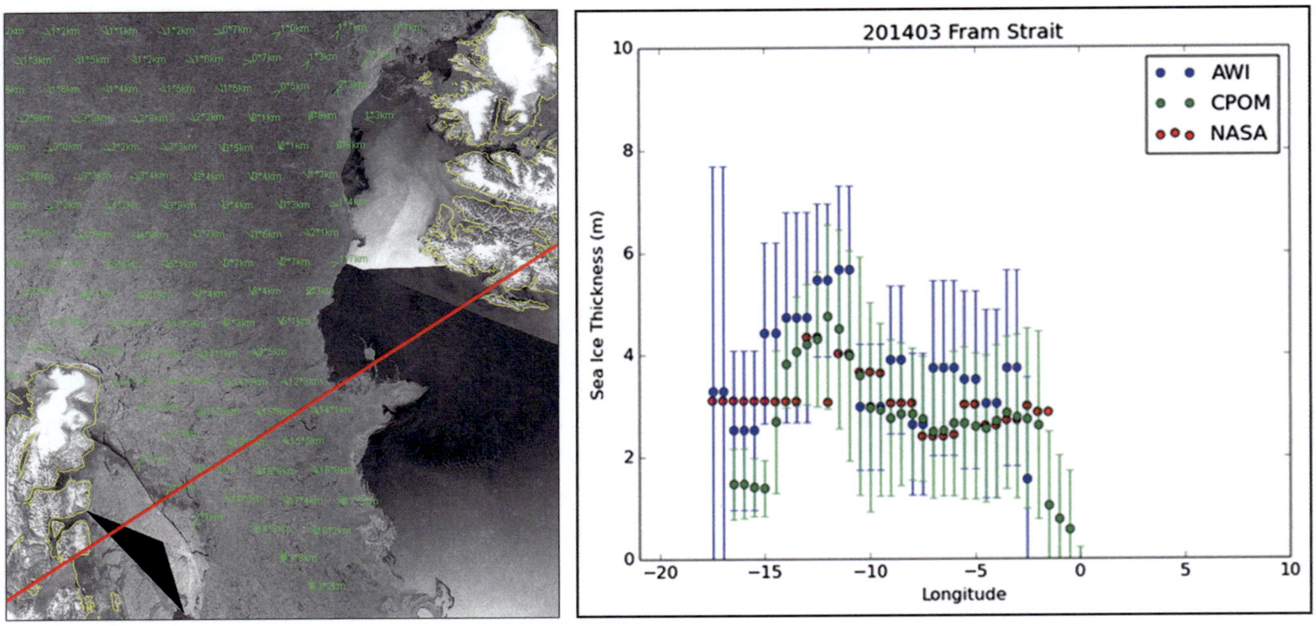

Fig. 6 Drift velocities from repeated S1A SAR imagery March 5th 2015 across the Fram Strait from seaice.dk (left). Red line represents the flux gate located at 79 N. (right) Sea ice thickness profiles from Fram Strait flux gate from 3 different CryoSat-2 derived products from March 2014

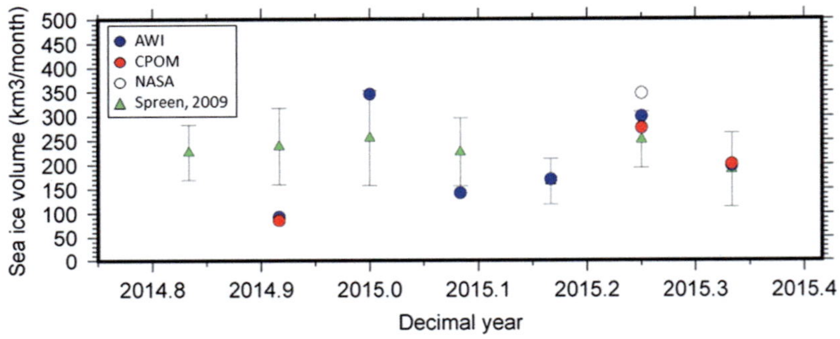

Fig. 7 Sea ice volume fluxes across the Fram Strait flux gate based on three Cryosat-2 based sea ice thickness profiles. Blue dots are from AWI, Red from CPOM and white from NASA. Green triangles are ICEsat estimates from Spreen et al. (2009) with uncertainty for comparison.

water fluxes have been calculated for the period October 2014 to April 2015 to prove the concept. We used daily drift vectors from feature tracking of repeated Sentinel-1 SAR-images accessed at seaice.dk across the Fram Strait located at 79°N as illustrated in the left panel of Fig. 6. Drift vectors were combined with monthly CryoSat-2 sea ice thickness products from three institutions; Alfred Wegener Institute (AWI), Center for Polar Observation and modelling (CPOM) and National Aeronautic and Space Administration (NASA) as shown in the right panel of Fig. 6. The AWI and CPOM estimates comes with error estimates. Even though the errors related with the sea ice thickness products are on the meter scale (vertical bars on Fig. 6 (left figure), the differences between the sea ice thicknesses from various products are within 0.5 m.

The monthly volume fluxes in this study (Fig. 7) show large variation with mean 216 km^3/month, minimum 86 km^3/month and maximum of 346 km^3/month. The large variation is primarily due to the highly variable sea ice drift on short time scales, daily and weekly, as these responds rapidly to changes in the wind field. A related study by Spreen et al. (2009) estimates sea ice volume through Fram Strait over the period 2002–2008 using ICESat data as input and daily drift vectors from passive microwave (AMSR-E). Spreen et al. (2009) found mean, minimum and maximum monthly estimates of 217 km^3/month, 92 km^3/month and 420 km^3/month. Our approach is within these limits, and demonstrates the potential to extend the time series including CryoSat-2 data. Compared with previous results using in-frequent ICESat data, CryoSat-2 allows to monitor continuous sea ice thickness for the first time. Combined with sea ice drift from repeated Sentinel-1A & B SAR-images continuous determination of sea-ice freshwater fluxes can be made. The high spatial resolution and daily coverage of the Sentinel-1A and 1B opens up for the possibility of estimating the sea ice fresh water flux even in the narrow Straits of the Canadian Arctic Archipelago (Fig. 1). It is further concluded from this study, that snow on sea ice only contributes with 2–7% of the total sea ice fresh water flux.

5 Ocean FWF

Ocean currents transport a considerable amount of fresh and saline water in and out of the Arctic Ocean (Proshutinsky et al. 2015). Satellite altimetry has for many years been applied to derive ocean currents under the geostrophic assumption through deriving an accurate mean dynamic topography. Until the launch of CryoSat-2, this has been an impossible task in the Arctic Ocean since it is partly ice covered. Hence, the new satellites are far less prone to be corrupted by sea ice and far better to capture sea level in leads of the sea ice (Armitage et al. 2016; Andersen and Piccioni 2016). In this study, the ocean fluxes were determined for the three major ocean flux gates: The Bering Strait (inflow) and the Fram and Davis Strait (outflow). Figure 8 below illustrates the Jason-1 tracks and the mean dynamic topography across the Bering Strait for monthly periods.

Assuming that the surface geostrophic flow is a good proxy for the depth mean velocity across the Bering Strait (Woodgate et al. 2015) the total water flux (F_T) can be determined from the slope of the monthly mean dynamic topography as illustrated in Cherniawsky et al. (2005).

The total amount of freshwater F_{FW} can be determined from the total water flux F_T using:

$$F_{FW} = F_T\ (1\text{–}SAL/REF)$$

Where the reference salinity (REF) for the Arctic Ocean is 34.8 PSU (Cherniawsky et al. 2005) and the monthly in-situ salinity observations (SAL) are from Woodgate et al. (2015).

Table 1 illustrate the seasonal variation in freshwater flux computed from satellite altimeter over 3-month periods. It highlights, that the freshwater flux is higher in the spring and summer month (April through September) and smaller in the autumn (October–December). The average annual freshwater flux from EO is 2,420 Gt/year from 6 years of Jason-1 data. This is in close agreement with (Serreze et al. 2006 and Haine et al. 2015) who found 2,400 Gt/year. For the Davis Strait the altimetric results were also within 5–7% of the results in (Serreze et al. 2006 and Haine et al.

Table 1 Transport and freshwater flux through the Bering Strait based on 6 years of Jason-1

	Transport	Freshwater Flux
Spring (April–June)	0.80 Sv	1980 Gt/y
Summer (July–September)	0.95 Sv	2,410 Gt/y
Autumn (October–December)	−0.27 Sv	−420 Gt/y
Winter (only January)	1.40 Sv	3,430 Gt/y
Average	0.95 Sv	2,420 Gt/y

Positive indicate a northwards transport and freshwater flux into the Arctic. Negative values correspond to outflow. Winter value was down-weighted as the peak only appears in January according to (Cherniawsky et al. 2005) due to the presence of sea ice

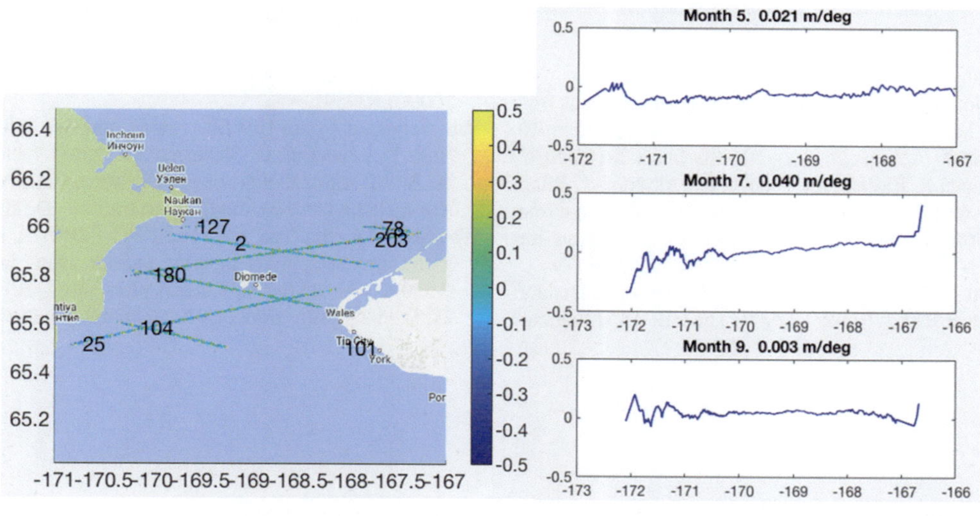

Fig. 8 (left) Jason-1 ground-track in the Bering Strait with pass numbers (track 180 used here). (right) Monthly averaged dynamic topography (relative to the EGM2008 geoid) for month 5, 7 and 9 (May, July and September) of 2006 and the computed linear slopes across the strait

2015). Altimetry enables computation of freshwater flux on monthly and even sub-monthly intervals since 1992 from the combined time series of TOPEX, Jason1 & 2 & 3.

In the Fram Strait (79°N) freshwater flux was determined from Cryosat-2. Here seasonal presence of sea ice degraded the agreement with the model studies to within 8–12%. Modelling Freshwater flux through the Fram Strait proved a bit more difficult due to a combination of sea ice and the inclination of Cryosat-2. However, the recently launched Sentinel-3A/B B satellites equipped with SAR altimetry will likely improve the results in this region and potentially enable monitoring of freshwater fluxes at sub-monthly interval.

6 Summary

In our view, Earth Observation data are essential to monitor freshwater fluxes and hence the freshwater budget in the Arctic Ocean on regular intervals. We have outlined the fundamental role of Earth Observations data in four cases studies (discharge from rivers; inflow through ice and melt run off; outflow of freshwater in sea ice and in/outflow of freshwater in the ocean). Our results generally demonstrate an agreement within 5–10% of the results in (Serreze et al. 2006; Haine et al. 2015) based on in-situ and modelling studies.

Acknowledgement The authors are thankful to ESA for funding the ArcFlux study under the STSE ITT Arctic+ program. The authors would also like to thank the space agencies and data repositories (RADS and seaice.dk) for providing state-of-the-art data for the investigation. Finally, the authors would like to thank an anonymous reviewer for valuable comments and suggestions to improve the manuscript.

References

Andersen OB, Piccioni G (2016) Recent arctic sea level variations from satellites. Front Marine Sci 3:76. journal.frontiersin.org/article/10.3389/fmars.2016.00076, https://doi.org/10.3389/fmars.2016.00076

Armitage TWK, Bacon S, Ridout AL, Thomas SF, Aksenov Y, Wingham DJ (2016) Arctic Sea surface height variability and change from satellite radar altimetry and GRACE, 2003–2014. J Geophys Res Oceans 121:4303–4322. https://doi.org/10.1002/2015JC011579

Bøssing Christensen O, Drews M, Hesselbjerg Christensen J, Dethloff K, Ketelsen K, Hebestadt I, Rinke A (2007) The HIRHAM Regional Climate Model. Version 5 (beta). Danish Climate Centre, Danish Meteorological Institute. Denmark. Danish Meteorological Institute. Technical Report, No. 06-17

Carmack EC, Yamamoto-Kawai M, Haine TWN, Bacon S, Bluhm BA, Lique C, Melling H, Polyakov IV, Straneo F, Timmermans M-L, Williams WJ (2016) Freshwater and its role in the Arctic marine system: sources, disposition, storage, export, and physical and biogeochemical consequences in the Arctic and global oceans. JGR Biogeosci 121(3):675–717. https://doi.org/10.1002/.JG003140

Cherniawsky JY, Crawford WR, Nikitin OP, Carmack EC (2005) Bering Strait transports from satellite altimetry. J Mar Res 63:887–900

Gallee H, Schayes G (1994) Development of a three-dimensional meso-γ primitive equations 5 model. Mon Wea Rev 122:671–685

Haine TW, Curry B, Gerdes R, Hansen E, Karcher M, Lee C et al (2015) Arctic freshwater export: status, mechanisms, and prospects. Glob Planet Chang 125:13–35

Kouraev AV, Zakharova EV, Samain O, Mognard E, Cazenave A (2004) 'Ob' river discharge from TOPEX/Poseidon satellite altimetry (1992–2002). Remote Sens Environ 93(1–2):238–245. ISSN 0034-4257. https://doi.org/10.1016/j.rse.2004.07.007

Morlighem M, Williams CN, Rignot E, An L, Arndt JE, Bamber JL, Zinglersen KB (2017) BedMachine v3: complete bed topography and ocean bathymetry mapping of Greenland from multi-beam echo sounding combined with mass conservation. Geophys Res Lett 44:11,051–11,061. https://doi.org/10.1002/2017GL074954

Nagler T, Rott H, Hetzenecker M, Wuite J, Potin P (2015) The Sentinel-1 Mission: new opportunities for ice sheet observations. Remote Sens 7:9371–9389

Nielsen K, Stenseng L, Andersen OB, Villadsen H, Knudsen P (2015) Validation of CryoSat-2 SAR mode based lake levels. Remote Sens Environ 171(15):162–170

Oltmanns M, Karstensen J, Fischer J (2018) Increased risk of a shut-down of ocean convection posed by warm North Atlantic summers. Nat Clim Chang 8:300–304

Proshutinsky A, Dukhovskoy D, Timmermans M-L, Krishfield R, Bamber JL (2015) Arctic circulation regimes. Philos Trans R Soc. https://doi.org/10.1098/rsta.2014.0160

RGI Consortium (2017) Randolph Glacier Inventory – A Dataset of Global Glacier Outlines: Version 6.0: Technical Report, Global Land Ice Measurements from Space, Colorado, USA. Digital Media. https://doi.org/10.7265/N5-RGI-60

Serreze MC, Barrett AP, Slater AG, Woodgate RA, Aagaard K, Lammers RB, Steele M, Moritz R, Meredith M, Lee CM (2006) The large-scale freshwater cycle of the Arctic. J Geophys Res Oceans (1978–2012) 111(C11)

Spreen G, Kern S, Stammer D, Hansen E (2009) Fram Strait sea ice volume export estimated between 2003 and 2008 from satellite data. Geophys Res Lett 36(19):L19502. https://doi.org/10.1029/2009GL039591

Van Meijgaard E, van Ulft LH, van de Berg WJ, Bosveld FC, van den Hurk B, Lenderink G, Siebesma AP (2008) Technical Report 302: the KNMI regional atmospheric climate model RACMO version 2.1. Royal Netherlands Meteorological Institute, De Bilt

Woodgate RA, Stafford KM, Prahl FG (2015) A synthesis of year-round interdisciplinary mooring measurements in the Bering Strait (1990–2014) and the RUSALCA years (2004–2011). Oceanography 28(3):46–67. https://doi.org/10.5670/oceanog.2015.57

Scientific and Operational Roadmap for Fiducial Reference Measurements in Satellite Altimetry Calibration & Validation

Stelios P. Mertikas, Craig Donlon, Rob Cullen, and Achilles Tripolitsiotis

Abstract

This work provides the essential elements for a scientific and operational roadmap with guidelines and practical directions to calibrate satellite altimeters under a new established standard of Fiducial Reference Measurements for Altimetry. According to this new principle, ground facilities, instrumentation and procedures, set up for calibration and validation (Cal/Val) of observations and products in altimetry, shall follow well-documented procedures and protocols, transparent to all involved. It shall also deliver uncertainty budgets of the Cal/Val results which should be built upon metrological standards and capable of being traced to Système International units. At first, this paper describes some guidelines for establishing new facilities for satellite altimetry Cal/Val with respect to their geographical location. Secondly, it gives requirements for maintaining an efficient performance and functionality for the Cal/Val facility. Third, it presents the optimal design and setup for the facility's instrumentation in relation to the Cal/Val technique employed. Finally, this work recommends in-situ data set formats.

Keywords

Altimetry · Calibration · Fiducial reference measurements · Roadmap

1 Introduction

Investigations and models for climate change progressively rely upon satellite products to develop but also validate seasonal to century projections of essential climate variables (Holman et al. 2013; Loew et al. 2017). Thus, satellite measurements and data products have to be homogeneous, stable, consistent, and continuous over a long period of time. Also, their uncertainty shall be capable of being traced to metrological standards and locked to undisputable SI (Système International) units (speed of light, atomic time, among others). To improve usefulness and reliability of climate data records as derived by satellites, the European Space Agency has established the concept of Fiducial Reference Measurements (FRM). This new FRM concept has been defined as: "*the suite of independent ground measurements that provide the maximum return-on-investment for a satellite mission by delivering, to users, the required confidence in data products, in the form of independent validation results and satellite measurements uncertainty estimation, over the entire end-to-end duration of a satellite mission*". Realization of this FRM standard takes place for several environmental parameters, such as satellite ocean color, air-quality, greenhouse gas observations, surface temperatures, optical time series, vegetation and sea level (ESA 2018).

This work provides a scientific and operational roadmap for achieving FRM quality in satellite altimetry (FRM4ALT) calibration and validation (Cal/Val). It is primarily based on

S. P. Mertikas (✉)
Geodesy & Geomatics Engineering Lab, Technical University of Crete, Chania, Crete, Greece
e-mail: mertikas@mred.tuc.gr

C. Donlon · R. Cullen
ESA/ESTEC, Noordwijk, The Netherlands

A. Tripolitsiotis
Space Geomatica PC, Chania, Crete, Greece

S. P. Mertikas, R. Pail (eds.), *Fiducial Reference Measurements for Altimetry*,
International Association of Geodesy Symposia 150, https://doi.org/10.1007/1345_2019_61

the experience gained after 15 years of continuous operation of the Permanent Facility for Altimeter Calibration (PFAC) in west Crete, Greece but also on international guidelines, best-practices, and recommendations for the future of satellite altimetry.

Today, there exist four absolute, permanent and historic such Cal/Val facilities in the world: One is operated by CNES (French Space Agency) in Corsica, France (Bonnefond et al. 2010, 2018) and elsewhere (Crétaux et al. 2018), One run by the Jet Propulsion Lab/NASA in California, USA (Haines et al. 2010), One managed by The University of Tasmania in Bass Strait, Tasmania, Australia (Watson et al. 2011) and one operated by the Technical University of Crete in west Crete and Gavdos in Greece (Mertikas 2018a, b, 2019).

Ground facilities, which are setup for satellite altimetry Cal/Val, shall be able to follow but also ready to adapt to new challenges and technological advances. These might include (a) New altimetry operating and measuring principles (i.e., Ka-band, Ku-band, SAR, nadir looking, interferometry, wide-swath, along-track), (b) Continuously raised scientific expectations and requirements for measuring sea-surface heights (i.e., mm-level for Sentinel-6), and (c) a changing global economic setting that prohibits large-scale investments and favors sharing among different space agencies the operational cost of Cal/Val facilities.

This paper presents an initial plan for fostering the transition of FRM4ALT methods and results (Mertikas et al. 2016), starting from research to operational activities, but also for identifying priority areas to be addressed right away. These include the geographical location for establishing new Cal/Val facilities for altimetry (Sect. 2), their efficient operation and performance assessment (Sect. 3), the instrumentation required to ensure delivery of uncertainty budgets for Cal/Val results following metrological standards (Sect. 4), Sect. 5 gives a proposal for the format with which Cal/Val site results will be handled. Finally, Sect. 6 lays out a summary roadmap along with strategies for integrating the FRM4ALT concept into existing initiatives and operations.

2 FRM4ALT Site Selection

Today, several complementary techniques have been implemented at various places on Earth (permanent, temporary) to calibrate various satellite altimeters. These Cal/Val sites have been principally located between the latitude zone of [−42°, +42°], indicating that large gaps need to be filled-in by new sites to improve global representation and Cal/Val reliability and efficiency (Fig. 1).

New sites may employ the sea-surface and/or a transponder calibration approach (Mertikas et al. 2018a), following the heritage of the existing Cal/Val sites. The availability of land and the associated permissions to host and operate the Cal/Val site are prerequisites for such an installation but will not be discussed further. The selection of candidate Cal/Val sites is recommended to rely upon a comprehensive set of criteria, such as for example: (a) multi-mission and cross-calibration capability at the same Cal/Val site. This

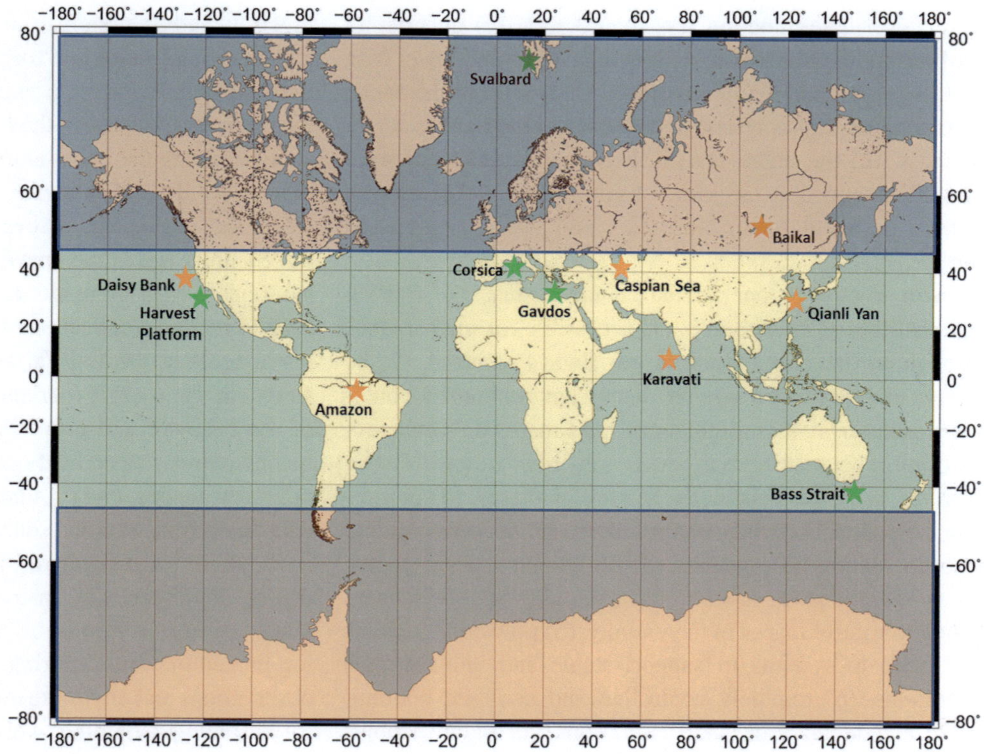

Fig. 1 Geographical distribution of permanent (green) and temporary (orange) satellite altimetry Cal/Val sites of the world

is a critical advantage as it permits comparisons and cross-evaluations of different missions against the same setting, and it increases the site's sustainability under the relatively sparse repeat cycle of current satellite altimeters (e.g., 10-days for Jason, 27-days for S-3A & -3B). The importance of this approach is highlighted by the choice of Copernicus Sentinel-3 orbit that was deliberately made to cross the PFAC in Crete, (b) the across-track (offset) distance of the site to satellite altimeter orbit shall be kept as small as possible, ideally ± 2 km, although, pending on conditions and the Cal/Val technique employed, larger distances (i.e., ± 15 km) may be accepted, (c) for sea-surface Cal/Val sites, signal contamination from land mass but also geophysical parameters (i.e., geoid, mean dynamic topography, ocean circulation, MSS, non-tidal effects) should be known precisely at the [cm] to [mm] level, (d) capability to support diverse Cal/Val techniques: sea-surface sites should make use of sea-surface but also crossover techniques, as transponder sites may engage range and sigma-0 calibration at the same location.

3 Facility Operational Performance Requirements

To calibrate altimeters properly, benchmarking parameters at the Cal/Val facility shall be determined with better accuracy than those given in the specifications for satellite observations. The following, as an illustration, are recommended as a minimum for reference measurements at the Cal/Val sites: (a) absolute positioning of the site better than 2 mm, (b) water level determination less than 3 mm, (c) uncertainty for transponder internal delay less than 30 ps and lower than 0.3 dB for its gain, (d) observations shall refer to the same, common coordinate system and relative to the Earth's center of mass, (e) the entire set of site observations shall refer to and be tagged at the same absolute time. Other aspects remain pertinent to specific sites and are not discussed further.

With respect to operational characteristics, the Cal/Val site shall be, as an example: (1) accessible for maintenance throughout the year (which is not always assured), (2) be secure and protected against harsh environmental conditions, vandalism, theft, lightening, (3) situated on stable, no-deforming bedrock/harbor, away from river runoffs, strong currents, with low seismicity, as well as others, (4) permitting tie connections between reference points on operating instrumentation (tide gauges, GNSS antennas, level points, transponder's phase center, for example) and permanent geodetic control marks in the vicinity, (5) powered by mains and back-up systems to support stable and uninterrupted operations, (6) remotely controlled and accessed continuously and at all times and conditions via secure, robust and redundant communications links. Warning flags for site conditions and failures shall be automatically issued and broadcast for actions, (7) unobstructed visibility to satellites, and permitting the installation of multiple instruments on the same location (see more in Mertikas et al. 2018a, b, 2019).

4 Instrumentation Requirements

The establishment of a satellite altimetry Cal/Val facility requires significant investment for infrastructure constructions, site development, installations and instrumentation. The least possible instrumentation for operating such a Cal/Val facility may include: (a) GNSS receivers to operate continuously and under internationally agreed standards (Bruyninx et al. 2017), (b) meteorological sensors to monitor, at least, ambient temperature, humidity, and atmospheric pressure, (c) water level measuring sensors (for sea-surface) or a microwave transponder (for transponder calibration), (d) reference clocks to specify the time at which this "benchmarking parameter" was used for establishing calibration, along with other metrological credentials.

Altimetry calibration requires ground measurements made by tide gauges, GNSS receivers, atmospheric sensors, oceanographic sensors, electronic monitoring devices, clocks, amongst others, where each of the contributing elements in the calibration process should itself be traceable using accepted metrology practices. The effort involved in establishing metrological traceability for each constituent in altimetry calibration should be commensurate with its relative contribution to the final measurement result, in precisely and accurately establishing an absolute sea-surface and/or ground reference for altimetry calibration. This is a demanding task and requires continuous assessment to assure the highest quality from the facility and its instrumentation.

To evaluate the uncertainty for the SI-traceable measurements (JCGM 2010) in the altimeter calibration, we have no means, at present, to revert to the absolute reference for the SI units [i.e., "the speed of light"] for establishing all subsequent measurements and their accuracy. At first, we have to rely on a collection of information for the calibrating instruments at the Cal/Val site, such as (1) previous measurement data, (2) experience with or general knowledge of the behavior and properties of relevant instruments, (3) manufacturer's specifications, (4) previous calibration or other certificates, and (5) uncertainties assigned to reference data taken from external sources, handbooks, and so on.

To attain FRM standards, as a practical approach we may deploy simultaneously instruments of different makes and kinds, diverse methods of measurement, various measuring procedures and differing approximations for best-fitting observations and environmental conditions. For example: (a) for absolute positioning, work with various GNSS receivers but also include other positioning systems, e.g., DORIS, Satellite Laser Ranging. It is recommended to install at least two multi-frequency, multi-constellation GNSS receivers but from different manufacturers, (b) for atmospheric signal delays, estimate delays not only by GNSS but alternatively by water vapor profilers, solar and lunar photometers, radiosondes, (c) for water level determination, operate more than three tide-gauges of different measuring principles (i.e., acoustic, pressure, radar, floating) and makes by diverse manufacturers, and so on. Consistency (or not) between measurements from different systems can be particularly helpful in understanding uncertainties from the different approaches.

Several uncertainties arise and finally contribute to delivered results for altimetry calibration. These may be associated with instrument type, measurement kind, measuring procedures and conditions, approximations, environmental conditions among others. Finally, an exhaustive statistical investigation should be contacted of every conceivable component of uncertainty and propagated to end products and thus describe the FRM uncertainty of final results for altimeter calibration. A critical tool that is used to guide and improve the PFAC has been the establishment and maintenance of a comprehensive uncertainty budget. From the start, the uncertainty budget analysis should be based upon the "Guide to the expression of uncertainty in measurement" (JCGM 2010).

5 Interface Control Documents

The operator of each permanent or temporary satellite altimetry Cal/Val facility shall disseminate and exchange the respective results in a transparent way using an internationally agreed format. Nonetheless, in-situ observations but also final results may be subject to intellectual property rights. Space agencies may however require an Interface Control Document (ICD) tailored to their specific needs and to their distinct satellite altimeters. Each Cal/Val site may produce its own primary ICD, dependent upon its particularities and characteristics, but a general ICD form that is maintained for all sites is highly desirable from a practical perspective. When an Agency requires additional information, then this ICD could be modified. It is suggested that the following principal and minimum components are to be included in each ICD of a Cal/Val site for uniform dissemination, namely: (1) Name and location of the Cal/Val site, (2) Year of establishment, (3) Satellite altimeters calibrated, (4) Type of calibration methods applied (sea-surface, transponder, crossovers, for example), (5) Reference systems and surfaces employed (reference system for coordinates, absolute time, ellipsoid, geoid, mean dynamic topography, solid Earth models), (6) Number, type, frequency of altimeter bias produced (e.g., radiometer bias, range bias, datation, sigma0, interferometric baseline length, space orientation), (7) FRM calibration values and their uncertainty budget with metrology standards applied, (8) Contact points, and so on. Ideally, a common Altimeter Cal/Val facility Web site that is used to exchange and disseminate information between operators and the end user community would be very beneficial. This has yet to be established, and possibly lined up with the metadata of GCOS (GCOS 2010).

6 Conclusions

There is an emerging need for standardization on Cal/Val products for satellite altimetry. The concept of Fiducial Reference Measurements for Altimetry adopts an SI standards based and fully traceable approach to fulfil this growing need for objective sea level change records. Under FRM4ALT, the scientific and operational roadmap addresses a strategy intended to achieve reliable, long-term, consistent and undisputable satellite altimetry products. In a succinct way such a roadmap should state, for example:

- Select a uniform, standard atomic time and coordinate reference system for all measurements used in the Cal/Val of satellite altimeters;
- Define a minimum set of essential observations, and ground-based instruments to support the Cal/Val;
- Establish a standardized way on how instruments (GNSS, tide gauges, transponders, meteorological sensors, radiometers) set up for defining "benchmark calibrating parameters", are to be characterized and calibrated before putting into use in the field;
- Institute the fundamental and undisputable metrology standards (e.g., light speed, atomic time) to build and place trust upon all measurements and results in altimetry calibration;
- Define error constituents (Mertikas et al. 2019), document all analytical procedures and practical steps to be followed for all FRM Cal/Val sites for describing and reporting uncertainty budgets for altimetry calibration;
- Put into operation procedures and techniques for evaluating differences in instrumentation measurements and for arriving at the "true" value of the parameter under investigation;
- Describe regular maintenance standards, following agreed protocols and characterization procedures (as an example, every six months tide gauges have to be sent to Lab for characterization in case deviations from a master tide gauge are noticed);
- Regulate the way of global distribution of Cal/Val sites with emphasis on strengthening weakness in altimetry by monitoring same orbits from diametrical Cal/Val sites on

the globe (verify the same errors from opposite sites on the globe);

- Establish a procedure for consolidated approach to data formatting, archiving and distribution, and
- Be prepared for the future of satellite altimetry Cal/Val, as new sites are to be ready to accommodate new measuring techniques. The Ka-band and Ku-band altimetry, two-dimensional and wide-swath altimetry are to become operational in a few years.

FRM4ALT is not about changing the way that existing Cal/Val facilities operate. It is about adding value and trust to the established procedures by evaluating in an objective way with traceability the uncertainty for their bias results. The determination of the uncertainty budget based on metrological standards, while a challenging exercise in it, constitutes the only undisputable indicator of the quality and reliability of the respective results. In a rapidly changing global environment, reconciling uncertainties in satellite based measurements is an essential scientific endeavor that adds credibility to the measurements used to monitor the impact of Government policies that increasingly impact our society.

Acknowledgements This work has been supported and funded by the EU and the European Space Agency (ESA No. 4000117101/16/I/BG).

References

Bonnefond P, Exertier P, Laurain O, Jan G (2010) Absolute calibration of Jason-1 and Jason-2 altimeters in Corsica during formation flight phase. Mar Geod 33(Suppl 1):80–90. https://doi.org/10.1080/01490419.2010.487790

Bonnefond P, Laurain O, Exertier P, Boy F, Guinle T, Picot N, Labroue S, Raynal M, Donlon C, Féménias P, Parrinello T, Dinardo S (2018) Calibrating the SAR SSH of Sentinel-3A and CryoSat-2 over the Corsica facilities. Remote Sens 10:92

Bruyninx C, Araszkiewicz A, Brockmann E, Kenyeres A, Legrand J, Liwosz T, Mitterschiffthaler P, Pacione R, Söhne W, Völksen C (2017) In: Villiger A, Dach R (eds) International GNSS Service 2017 Technical Report. IGS Central Bureau and University of Bern; Bern Open Publishing, pp 105–115. https://doi.org/10.7892/boris.116377

Crétaux J-F, Bergé-Nguyen M, Calmant S, Jamangulova N, Satylkanov R, Lyard F, Perosanz F, Verron J, Samine Montazem A, Le Guilcher G, Leroux D, Barrie J, Maisongrande P, Bonnefond P (2018) Absolute calibration or validation of the altimeters on the Sentinel-3A and the Jason-3 over Lake Issykkul (Kyrgyzstan). Remote Sens 10:1679

European Space Agency (2018) Fiducial Reference Measurements: FRM. ESA Sensor Performance, Products and Algorithms. https://earth.esa.int/web/sppa/activities/frm. Accessed 26 July 2018

GCOS (2010) Guide to the GCOS Surface Network and GCOS Upper-Air Network, Global Climate Observing System, November 2010, GCOS–144, WMO/TD No. 1558

Haines BJ, Desai S, Born G (2010) The Harvest Experiment: calibration of the climate data record from TOPEX/Poseidon, Jason-1 and the Ocean Surface Topography Mission. Mar Geod 33(Suppl 1):91–113. https://doi.org/10.1080/01490419.2010.491028

Holman R, Merchant CJ, Saunders R, Downy C, Buchwitz M, Cazenave A, Chuvieco E, Defourny P, De Leeuw G, Forsberg R, Holzer-Popp T, Paul F, Sandven S, Sathyendranath S, Van Roozendael M, Wagner W (2013) The ESA climate change initiative satellite data records for essential climate variables. Bull Amer Meteor Soc 94:1541–1552. https://doi.org/10.1175/BAMS-D-11-00254.1

JCGM (2010) Evaluation of measurement data Guide to the expression of uncertainty in measurement. Bureau International des Poids et Mesures. https://www.bipm.org/utils/common/documents/jcgm/JCGM_100_2008_E.pdf. Accessed 30 July 2018

Loew A, Bell W, Brocca L, Bulgin CE, Burdanowitz J, Calbet X, Donner RV, Ghent D, Gruber A, Kaminski T, Kinzel J, Klepp C, Lambert J-C, Schaepman-Strub G, Schröder M, Verhoelst T (2017) Validation practices for satellite-based Earth observation data across communities. Rev Geophys 55:779–817. https://doi.org/10.1002/2017RG000562

Mertikas S, Donlon C, Mavrocordatos C, Boikov B, Femenias P, Parrinello T, Picot N, Desjonqueres JD, Andersen OB (2016) A fiducial reference site for satellite altimetry in Crete, Greece. Poster, Ocean Surface Topography Science Team Conference, 1–4 November, 2016, La Rochelle, France

Mertikas SP, Donlon C, Féménias P, Mavrocordatos C, Galanakis D, Tripolitsiotis A, Frantzis X, Tziavos IN, Vergos G, Guinle T (2018a) Fifteen years of Cal/Val service to reference altimetry missions: calibration of satellite altimetry at the permanent facilities in Gavdos and Crete, Greece. Remote Sens 10:1557

Mertikas S, Donlon C, Féménias P, Mavrocordatos C, Galanakis D, Tripolitsiotis A, Frantzis X, Kokolakis C, Tziavos IN, Vergos G, Guinle T (2018b) Absolute calibration of the European Sentinel-3A surface topography mission over the permanent facility for altimetry calibration in west Crete, Greece. Remote Sens 10(11):1808

Mertikas S, Donlon C, Femenias P, Cullen R, Galanakis D, Frantzis X, Tripolitsiotis A (2019) Fiducial reference measurements for satellite altimetry calibration: the constituents. Fiducial reference measurements for altimetry, Proceedings of the international altimetry Cal/Val review & applications. Springer, Heidelberg. https://doi.org/10.1007/1345_2019_56

Watson C, White N, Church J, Burgette R, Tregoning P, Coleman R (2011) Absolute calibration in Bass Strait, Australia: TOPEX, Jason-1 and OSTM/Jason-2. Mar Geod 34(3–4):242–260. https://doi.org/10.1080/01490419.2011.584834

List of Reviewers

Baki Iz
C. Hwang
Christine Hackman
Constantin Mavrocordatos
Demitris Delikaraoglou
Denise Dettmering
Ernst Schrama
Georgios S. Vergos
Gerard Petit
Jesus Gomez-Enri
Laurence Padman
Luciana Fenoglio-Marc
Martin Jakobsson
Neil Ashby
Nikita Zelensky
Ole Baltazar Andersen
Philippe Schaeffer
Ronald Kwok
Stelios Mertikas
Vassilis Gikas
Xiaoli Deng

S. P. Mertikas, R. Pail (eds.), *Fiducial Reference Measurements for Altimetry*,
International Association of Geodesy Symposia 150, https://doi.org/10.1007/978-3-030-39438-7

Author Index

S. P. Mertikas, R. Pail (eds.), *Fiducial Reference Measurements for Altimetry*,
International Association of Geodesy Symposia 150, https://doi.org/10.1007/978-3-030-39438-7